AFFORDABILITY
Integrating Value, Customer, and Cost for Continuous Improvement

Continuous Improvement Series

Series Editors:
Elizabeth A. Cudney and Tina Kanti Agustiady

PUBLISHED TITLES

Affordability: Integrating Value, Customer, and Cost for
Continuous Improvement
Paul Walter Odomirok, Sr.

Design for Six Sigma: A Practical Approach through Innovation
Elizabeth A. Cudney and Tina Kanti Agustiady

AFFORDABILITY

Integrating Value, Customer, and Cost for Continuous Improvement

PAUL WALTER ODOMIROK, SR.

CRC Press
Taylor & Francis Group
Boca Raton London New York

CRC Press is an imprint of the
Taylor & Francis Group, an **informa** business

CRC Press
Taylor & Francis Group
6000 Broken Sound Parkway NW, Suite 300
Boca Raton, FL 33487-2742

Printed by CPI on sustainably sourced paper
Version Date: 20161102

International Standard Book Number-13: 978-1-4987-6240-3 (Hardback)

Library of Congress Cataloging-in-Publication Data

Names: Paul Walter Odomirok, Sr., author.
Title: Affordability: Integrating Value, Customer, and Cost for Continuous
    Improvement / Paul Walter Odomirok, Sr.
Description: Boca Raton : Taylor & Francis, a CRC title, part of the Taylor &
Francis imprint, a member of the Taylor & Francis Group, the academic
division of T&F Informa, plc, [2016] | Includes index.
Identifiers: LCCN 2015038582 | ISBN 9781498762403
Subjects: LCSH: Engineering mathematics. | Manufacturing
processes--Mathematical models.
Classification: LCC TA330 .P39 2016 | DDC 620.0068/4--dc23
LC record available at http://lccn.loc.gov/2015038582

Visit the Taylor & Francis Web site at
http://www.taylorandfrancis.com

and the CRC Press Web site at
http://www.crcpress.com

Contents

Preface

The concept of affordability emerged in 2007 during the early part of a project Dr. Elizabeth Cudney and I were participating in at a major defense aerospace production facility in California. That plant, even though it produced an excellent product, was scheduled to close in August of 2010, and its role at that time was only to fulfill the product replacement needs of its customer. We were there to incorporate Lean methods and behaviors for improvement as part of an ongoing corporate endeavor of continuous improvement. The only customer for this product was the U.S. Department of Defense, and the product was an advanced version of a defense aircraft that was first released in 1978. During the early days of the project, we discovered that, although Lean implementation efforts had been attempted in 2000, 2002, and 2004, with each try failing, we had to set an effective aim for continuous improvement that would focus and motivate the workforce, serve as a target for the alignment of improvement efforts, and take the organization beyond its limited, short-term horizon. The project and program became known as "Affordability." It provided a purpose and target consisting of the primary components of value, customer, and cost. A few years later, by August of 2010, the product had several customers, the unit cost was reduced by $5.1 million dollars, and the U.S. Department of Defense ordered 124 more aircraft, saving the American taxpayer more than $600 million, extending the life of the facility until 2020, protecting jobs, and ensuring ongoing organization success. The then assistant secretary of defense, Dr. Ashton Carter, distributed a memo on September 14, 2010, to all acquisition personnel professionals entitled "Better Buying Power: Guidance for Obtaining Greater Efficiency and Productivity in Defense Spending," where on page 7, paragraph 3, he refers to this product as an example of how to gain better buying power. From this event, the concept, theory, and philosophy of affordability were born.

Today, many leaders and managers struggle, and are frustrated with, implementing continuous improvement. Much of their exasperation is caused due to the fact that "continuous improvement for the sake of continuous improvement" does not stick nor resonate with the people and culture of the organization. Continuous improvement has no aim. Affordability is the effective aim continuous improvement has lacked. This book provides individuals with a foundation and framework of how to go about instituting continuous improvement through the aim of Affordability. It provides examples of how others have done it. It supplies tools and toolboxes that serve to create solutions and fix problems. And, it outlines the primary components and pertinent factors for achieving affordability.

Across the world today, we've experienced a notable shift in the global economy, especially with the traditionally strong industrial nations of the United States, Japan, and Germany. Some of these economic adjustments have occurred because the clarity of the elements of value, customer, and cost has been blurred by an effort to produce faster, better, and cheaper. Affordability addresses all three of these dimensions and provides an approach to satisfy them. I've tested the concept and theory of affordability with organizations in North America, the Middle East, and Europe. Continuous improvement is founded upon faster, better, and more affordable.

For me, I've been on the "journey of affordability" for a long time. After spending almost 10 years in academia, a little more than 10 years in corporate America with NCR and AT&T, and another 20 years consulting and serving more than 80 organizations, I've realized that a key factor in longevity is perseverance and continuous improvement. Affordability is the aim for continuous improvement through a trifocal lens. It's all about focus on delivering value, on providing for the customer, and on the everlasting pursuit of reducing cost and providing more affordable products and services. Although it's taken a long time for compiling all the examples and evidence, for discovery, for documenting the concepts, and for this journey to come to fruition, it's now time for others to realize benefits from establishing affordability as the aim and goal for everyone to pursue.

Where value and customer intersect, we identify the requirements, and the "how to satisfy the customer in terms of the 'what,' the 'when,' and the 'how much.'" Where customer and cost intersect, we have the established price point of what the customer is willing to pay, as well as other value points for the customer. Where cost and value intersect, we realize the expense of what it takes to deliver value to the customer. Where they all come together, this is the AIM, the point of integration, the triumphant center of success. The AIM is achieved when the right products and services are provided to the customer at the right price, under the right cost, meeting and exceeding the customers' expectation. When at NCR, my triple AIM was banking and retail products that met and exceeded customer needs, at a reasonable price provided through a cost and expense structure that kept the organization profitable. In manufacturing, providing products that customers are willing to purchase from profitable organizations, at a reasonable price, achieves the AIM. In healthcare, the AIM is healthy communities (value), positive patient outcomes (customer), at lower expense and price. In any organization, serving customers, providing products and/or services, the AIM of affordability provides a framework and focus for success.

As your read this book, you'll likely realize, as I did, that affordability has always been with us. However, until now, its context and content have not been well defined and applied in a way to advance any type of organization toward realization of that ever-elusive quest of excellence through continuous improvement. In each chapter, I've included examples of when affordability is present and when it is absent. Performance excellence and sustained success are realized through growth and longevity, and affordability serves as the basis for such results and outcomes.

Author

Paul W. Odomirok is president and CEO of Performance Excellence Associates, Inc., Lawrenceville, Georgia. For the past 40 years, Paul has been involved in several careers from academia to corporate leadership to consulting to entrepreneurship. In his "first career," he was a mathematics and computer science instructor at all levels of learning, from preschool to post-PhD. He was even involved in designing graduate-level curriculums at the University of South Carolina as an adjunct professor.

His second career began in 1985 with NCR. Beginning as a senior programmer analyst, he was promoted to Manager—Software Development, Product Manager, Manager—Product Management, Corporate Strategic Planner for Banking, Director—Retail Systems Product Integration, Director of Quality, and Corporate Coach. During his stint in corporate America, he experienced the NCR/AT&T merger and was responsible, as a director of quality, for the cultural transformation and change leadership for the Retail Systems Division Organization in Duluth, Georgia. He was trained by Bell Labs and utilized as an executive coach for NCR/AT&T executives.

He left NCR/AT&T in 1995 to pursue a consulting career in the areas of leadership, team development, strategy, structure, systems, and organization performance. Over the past 20 years, he has trained, coached, and mentored hundreds of IIE and ASQ Green Belts and Black Belts in Lean, Six Sigma, and Lean Six Sigma, as well as coached leaders and managers on how to develop and implement strategic plans and organization transformation programs. He's worked with more than 80 organizations on more than 160 different performance improvement projects, organization change initiatives, and continuous improvement programs. His expertise ranges across a variety of disciplines, including Lean, Six Sigma, manufacturing, engineering, supply chain, healthcare, multiple services, and leadership/management development, and a multitude of industries. Although most of his projects have been concentrated in the commercial business area, his most recent projects have been focused on national defense and healthcare for leadership development, strategic planning, increasing process speed, improved quality, affordable cost, and supply chain logistics excellence for specific programs such as the F/A-18 "Super Hornet" and the MRAP (Mine Resistant Ambush Protected) vehicle program and for several healthcare institutions.

Both his BS and MS are concentrated in mathematics, which he utilizes today for complex problem solving, implementing transformation and change, project management, Lean Six Sigma Green/Black Belt training through the Institute of Industrial and Systems Engineers, Lean principles incorporation, process statistics utilization, and other fact and data analysis methods and approaches. He has been involved in research with Bell Labs for technical team design and served on a Harvard research team called "The Events and Motivation Study" (the HBS TEAM Study; for research results, see "The Progress Principle" by Dr. Teresa Amabile and Dr. Steven Kramer).

chapter one

Affordability

It's not what we always thought it was!

> I will build a car for the great multitude. It will
> be large enough for the family, but small enough
> for the individual to run and care for. It will be
> constructed of the best materials, by the best men
> to be hired, after the simplest designs that modern
> engineering can devise. But it will be so low in
> price that no man making a good salary will be
> unable to own one – and enjoy with his family the
> blessing of hours of pleasure in God's great open
> spaces.
>
> **—Henry Ford**

It was more than a hundred years ago, Henry Ford, using strong wisdom and a sound business philosophy, created the initial foundation and framework for affordability. Affordability is traditionally defined as "being within one's financial means." However, affordability as related to continuous improvement is an aim and, as a point of integration, is focused on increasing value, improving customer focus, and decreasing cost. If a product or service of an organization is of value, meets the needs and wants of the customer, is sold at a cost within the customers' means, and is profitable for the provider organization, it meets the core definition of affordability, and the three points of focus can be leveraged for continuous improvement efforts over time. When used effectively, affordability becomes the primary target and prime objective for an organization to constantly pursue excellence and chase perfection.

Affordability is a value-centered, customer-focused, cost-effective approach to provide products and services with correct positioning in the target market and marketplace; with correct profitable pricing; containing a value proposition that meets and exceeds customers' expectations, needs, wants, and requirements; that is delivered as fast as required; and containing excellent quality and outstanding reliability. Over time, affordability drives higher value, greater availability, and continuous improvement in quality and reliability at a price that appropriately fits the target market, with cost being continually reduced over time. Products and services within this construct often maintain, and even lower, their price with improved versions, while adding greater features and value at little or no additional cost increase to the customer in initial pricing or reliability expenses. Products and services that adhere to the affordability framework are sustainable and continue to be economically feasible in the customer target area for long periods of time. Affordability achieves customer loyalty as it helps to grow and expand an organization's market share.

Affordability provides direction, setting the path for an organization's pursuit of continuous improvement. As is often the case, a corporation or institution desires continuous improvement; however, it does not have a business case focal point for such endeavors. Over the past 40 years, I've encountered "quality," "zero defects," "lean," "elimination of waste," "Six Sigma," "variation reduction," and even "continuous improvement," as well as many other nuclei for focusing continuous improvement. However, the emergence of a formidable aim for continuous improvement has not appeared. One of the first few steps in any substantial change or transformation effort is the establishment of a direction.

Affordability is also about alignment. It's about the alignment of value, customer, and cost all in an effort to collaboratively increase effectiveness. With this alignment, it guarantees a degree of capability for customers to purchase a sustainable product or service that meets their requirements, needs, wants, and desires. Affordability alignment also promises that for the life of the product or service, reasonable costs for maintenance and sustainment exist.

Affordability is also about integration. It's about the integration of value, customer, and cost for continuous improvement. Increasing value, exceeding customer expectations, and reducing cost continually is a straightforward, simple approach for achieving continuous improvement. Organizations utilizing affordability as a strategic platform can achieve market share growth, competitive advantage, and organizational success.

Since 1995, I've served more than 80 organizations. Sometimes by the way of projects for improvement or leadership, at other times via knowledge transfer and training, and even as a result of coaching and mentoring. In all the services I've provided, when the value, customer, and cost were aligned, the project, training, coaching, and mentoring occurred. When the point of integration didn't align, I didn't get the job. I've observed this same phenomenon across the various industries I've been involved with, and the numerous customers I've served.

This is the model of affordability (Figure 1.1). The three primary components of affordability are value, customer, and cost. When aligned, these factors designate and illustrate a balance of the aspects that achieve affordability.

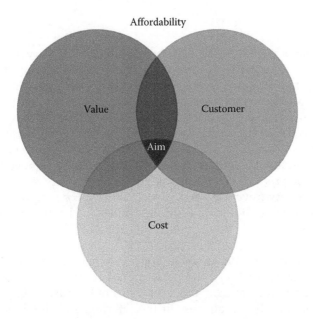

Figure 1.1 The model of affordability.

Value, customer, and cost, when detailed and defined, provide focal points for every organization to center on and emphasize for success. Each component can be summarized as follows:

Value

- Value is defined as "what you provide that the customer requires, wants, and needs, and is willing to pay for."
- Value is the importance of the organization, what has worth for the customer, what is useful, what is needed, what is wanted, and what is required of a product or a service from the customers' perspective.

Customer

- Customer is defined by those who purchase, those capable of purchasing, those recipients of the organization's work, and those targeted for purchasing the product or service being offered.
- The customer, or individuals within the targeted market, is the aggregate of buyers or purchasers or recipients of the product or service being offered.

Cost

- Customer cost is defined as the customer's actual overall cost of a product or service, or total operating cost including the purchase price, maintenance fees, associated costs, and any other affiliated costs related to that product or service.
- Organization cost is viewed from the organization's perspective, with consideration of all costs, including direct costs, labor costs, and indirect costs, to produce or provide the product or service using an internal viewpoint. When organization cost is removed from the revenue produced from the sale of a product or service, the result is profitability (revenue − cost = profitability). For nonprofit organizations, when cost equals funding available, a balanced condition is achieved.

The area of intersection of value and customer defines requirements, wants, needs, expectations, and performance criteria. This zone provides knowledge and information about the voice of the customer, the market, the competition, the uniqueness of the offering, and the basic template elements of what it takes to be successful in the business (Figure 1.2).

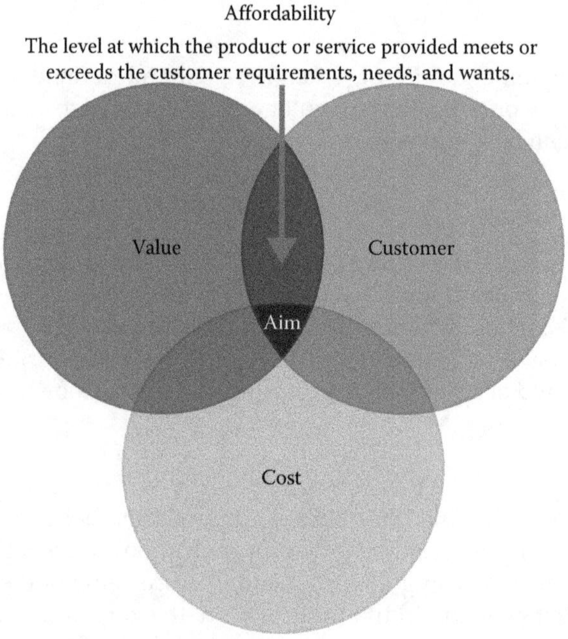

Affordability
The level at which the product or service provided meets or exceeds the customer requirements, needs, and wants.

Value

Customer

Aim

Cost

Figure 1.2 The intersection of value and customer primarily involves customer requirements and value delivery.

The place where value and cost come together portrays the investment and expense necessary to achieve the purpose and intent of product and/or service the organization is providing. It encapsulates the direct expenses, the indirect expenses, the variable expenses, the cost of quality (and the cost of nonquality), and the required success and performance measures, along with the ability and capability to solve problems and resolve issues (Figure 1.3).

Where customer and cost overlap, price or customer cost is established. This ranges from the initial financial outlay for the customer, to the ongoing maintenance and sustainment charges and fees, to the benefits attained and retained by purchasing the artifact or amenities being purchased (Figure 1.4).

The overall aim, or the meeting of all three components, articulates the requisite for enhancing value, exceeding customer expectations, decreasing cost, growing market share, and increasing demand. It provides the platform for every organization to establish their own aspirations of excellence. It is the aim, the point of integration, and the area of alignment. For system sustainability and the endless pursuit of continuous improvement, this aim for alignment and integration is the place where balance and stability can be measured, monitored, and managed. When off balance, or off kilter, the system subsequently becomes unbalanced and unstable. I've witnessed this in corporations where value, customer, and cost do

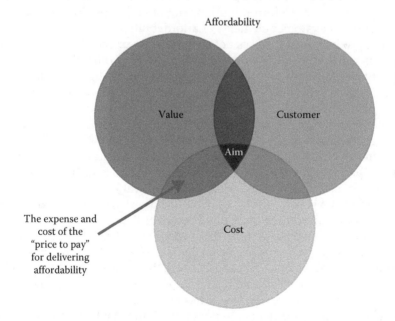

Figure 1.3 Expense is the primary factor of the intersection of value and cost.

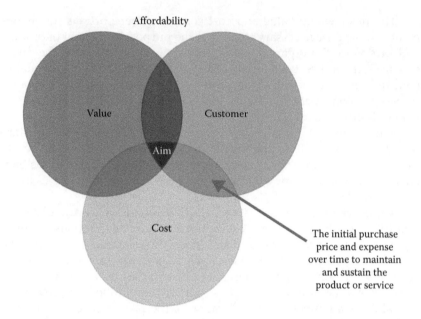

Figure 1.4 Price is the primary factor of the intersection of customer and cost.

not align and do not integrate in a balanced and stable manner. When costs get out of control, from both operational and pricing perspectives, the value and customer elements of affordability suffer. Customers cease buying products and services not competitively priced. The value proposition is ruined when costs exceed revenues and profitability approaches (or achieves) a negative zone. When value is out of alignment and stable with customer, cost also suffers attempting to provide value that does not align with customer requirement and expectations. Another good contemporary example for misaligned and unstable conditions for affordability is government. In the Preamble of the Constitution of the United States (September 17, 1787): "We the People of the United States, in Order to form a more perfect Union, establish Justice, insure domestic Tranquility, provide for the common defense, promote the general Welfare, and secure the Blessings of Liberty to ourselves and our Posterity, do ordain and establish the Constitution for the United States of America." With this value proposition and purpose, over 228 years have elapsed, and numerous laws and policies have been enacted. In its current state, at the time of this writing, costs have considerably increased, the value proposition has been enlarged beyond its basic intent, and the customer base, U.S. citizens, has grown dramatically, changed in terms of requirements, and as a result, the three affordability elements are not aligned and stable with the original intent and reason for the founding of the country. Hence, using facts only,

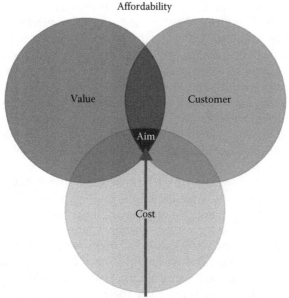

The aim of affordability is to increase value, exceed customer expectations, and
reduce the costs that affect the products and services through continual
process improvement and continuous organizational improvement

Figure 1.5 The aim of affordability is the triumphant balance of value, customer
and cost including requirements, delivery, expense, and price.

and not politics, the affordability of the U.S. government is misaligned
and unstable. This has been clear over the past 20 years as the value is
being changed, the customer base is being changed, and a great deficit
budget has accumulated. The solution is to reestablish the aim around
value, customer, and cost. This is to not establish the aim on any one or
two of the elements, but all of the elements with the same emphasis and
level of importance (Figure 1.5).

Recently, I've been able to witness affordability firsthand as a result
of the multiple projects that I was involved in with the Mine Resistant
Ambush Protected Program from August 22, 2007, until March 15, 2014.
During this time, the value in the program was established as protecting the lives of soldiers, now commonly referred to as warfighters, from
improvised explosive devices in war on the Afghanistan and Iraq theaters of battle. The definition and requirements for the customers of the
program were clearly established by the Department of Defense, and their
needs, wants, and wishes were distinct and specific. From a cost perspective, the program was funded at the expected spend for $55 billion, while
the final overall expenditure was $52 billion (note: $3 billion went unspent

and was returned to the government). Each vehicle was priced at about $1.2–1.4 million (including logistics and distribution, and maintenance and sustainment material), and the operations of the damage and repair systems were provided as part of the funding. The deployment and operation of the program lasted from 2007 to 2012, and the program transferred the vehicle responsibilities to the various services in 2013. All in all, the value of the program advanced and increased over the program horizon, the customer base grew and matured from a U.S.-only initial focus to a multinational focus for all allies involved, while the expected cost and expense was decreased and reduced from the initial estimates and expectations (Figure 1.6).

Looking farther back in time, the basics and groundwork for affordability is more than 100 years old. Since the beginning of the twentieth century, affordability has been commonly used within the contexts of housing, healthcare, automobiles, energy, higher education, transportation, insurance, textbooks, along with many other environments, and under many other conditions worldwide. Affordability for an individual is defined by the capability and ability of that individual to purchase and acquire a product or service for their own use and consumption, in accordance with their wants, needs, expectations, and requirements. Affordability in any market is also traditionally defined by the capability of the customers in that market grouping to purchase and maintain the purchased offerings.

In today's marketplace, affordability is often confused with the concepts of "cheap" and "free," and is frequently abused and misused to convince people to purchase a product or service that does not fit within a

MRAP (mine-resistant ambush-protected) vehicles

- Purpose/intent: Save lives, mitigate
                         IED deaths and injuries
- Integration   8/2007→11/2007
                         5/day→50/day
- 3-year deployment   27,500 (*fastest ground combat system fielding ever*)

- "The realized value"
    - 6 warfighters   @ $500,000/1 Warfighter X 6 = $3M
    - Lives protected   @ $3M/vehicle ~ $82.5BB lives protected
    - Vehicle cost   ~ $38,500,000,000
    - Maintenance   ~ $13,500,000,000
    - Duration   FY07 – FY14
    - Program cost $52BB ($3BB returned to govt.)
                    Overall value: $$$ billions

Figure 1.6 Mine Resistant Ambush Protected Vehicle Program—MRAP.

reasonable cost or price range that resides within their ability to purchase. Recently, there has been a debate in America about "affordable healthcare." True affordable healthcare is provided at an overall cost that can be acquired at a reasonable price, meeting the needs, wants, and requirements of the individual receiving care, with positive patient outcomes.

Affordability is not limited to products and services alone. From an enterprise perspective, the affordability model can be applied to systems for increasing efficiencies and effectiveness. Such systems, in addition to providing products or services, can also offer a support infrastructure for continuously improving finance, operations, production, information technology, engineering, sales and marketing, and other structural entities within a corporation or establishment. The "value" is defined as their functional purpose, while their customers are the individuals whom the delivery services to, and their costs are the financial expenses and fees that are required. For example, I've been involved in several healthcare projects over the past 20 years. Recently, within this marketplace, a version of affordability has emerged and is being recommended by the Institute for Healthcare Improvement (IHI) as "Triple Aim" for healthcare organizations to optimize health system performance. It is focused on value → *healthy communities*, customer → *patient outcomes*, cost → *healthcare costs and expenses*. Although "the payer" in this particular market is primarily insurance companies, the customers in this market must directly purchase healthcare insurance or they are provided healthcare insurance through their workplace or by the government. Of course, there are many combinations and permutations of this formula. However, a universal truth exists; it is not cheap, it is not free, but it does conform to what is required, and it can be purchased within the financial means of the buyer. Currently, the alignment of value, customer, and cost has not clearly emerged. Time will tell whether this approach will reach affordability alignment and prove to be right for the American public. For the entire population, there are a multitude of healthcare needs and market segments, and in defining affordability, it must be taken into account the value–health communities, the customer–patient outcomes, and the cost–cost of healthcare services. Affordable healthcare's target, as defined by the IHI's "Triple Aim" is focused on healthy communities ("population health"), patient outcomes (experience of care), and healthcare cost (per capita cost). This model applies to the individuals served in this market, as well as the systems involved in this marketplace. For more information, see: www.ihi.org/engage/initiatives/tripleaim/Pages/default.aspx.

In local regional and federal government, another environment I've been involved with over the past several years, affordability is framed by value—*community and citizen services*; customer—*neighbors or citizens*; and cost—*tax money applied to the services, operations, and programs provided to the population.* In yet another government-related instance I've worked with, Defense Aircraft Manufacturing, the foundation of affordability is based on value—*high-speed effective military aircraft or commercial aircraft,* customer—*U.S. and U.S. allies, or airline service companies,* and cost—*at an established initial price with sustainable maintenance costs with reliability and durability for dozens of years.* And finally, another example of government affordability came from a Department of Defense project I mentioned earlier based on value—*protecting lives,* customer—*American warfighters and American allies,* and cost—*at program end, the overall cost came in at a level of around 95% of the amount congress budgeted for in October of 2006 that returned a few billion dollars back to the federal budget (a government project on time and below budget!).*

Beyond products and services viewpoint, affordability as viewed from the enterprise and systems perspective, one must consider the priority of systems or hierarchy of functions for a prioritized approach. Since affordability is customer centered, the affiliated systems and functions delivering products and services begin with those systems that directly provide customer care and support, and transition down through all services and functions that support the value-added areas, referred to as value-added support, as well as those systems that are in place to support the entire organization from leadership and management throughout each establishment facet. Figure 1.7 illustrates the affordability hierarchy of systems.

There are seven layers of systems and system processes for functions that exist within the affordability hierarchy. In smaller organizations, individuals perform many different functions and the formidable distinction of these layers ceases to exist. As organizations grow larger,

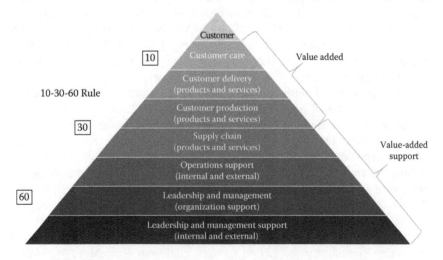

Figure 1.7 Systems hierarchy.

the functions of each layer become distinct, and function specialization at each tier tends to crystallize, and the tendency for "silos" or "stove pipes" to appear increases. Large organizations must constantly strive to keep the systems integrated and focused on the value and customer of the organization.

Customer care: This consists of the functions and processes that directly serve the customer. In a manufacturing environment, customer care serves the customer to assist the customer in procuring the organization's product(s) and also to support the organization's product at the customer site. In a service organization, customer care directly provides the service and assists in supporting the service provided.

Customer delivery: An integrated set of processes and procedures, along with the necessary functions to bring the product or service to the customer. Customer delivery may be configured such that the customer must come to an established site where the customer can acquire the product or service (e.g., retail store, service center, a designated organization location). Customer delivery may also be provided via logistics methods configured to bring the product or service to the customer. Or, customer delivery may occur through means of a third-party supplier delivering the product or service.

Customer production: Whether an organization provides product(s) and/or service(s), the value for the customer exists in the production of a product or service available and accessible for the customer to obtain. Production usually implies product, but in this case, production also infers service due to the requirement that a service requires preparation, fabrication, and assembly of the components of the service for delivery to the customer.

Supply chain: The components, materials, methods, and means for a product or service must be available for production and delivery. The supply chain is the integrated linkage of all the processes and procedures needed for production and delivery.

Operations support: There exists a multitude of functions and processes required to support the operational components of an enterprise. For any product or service to be produced and delivered, the unencumbered flow of information, material, and support that facilitate efficient and effective production and delivery is necessary. All functions and systems that support such flow fall into the category of operations support.

Leadership and management: The leadership of people and the management of nonhuman resources maintain a continuity for any organization to thrive. The purpose of leadership and management within the affordability framework is to set direction, align the resources, motivate the people, communicate and execute the plan and strategy, and maintain an environment of success for the people, processes, and performance of the organization.

Leadership and management support: The organization functions that support the success of leadership and management, as well as the production and delivery of customer value fall into this category (e.g., finance and accounting, legal, human resources, information technology, sales and marketing, engineering, manufacturing, sourcing and purchasing, quality, service and support, facilities, and product and service management).

While collaborating on a project with Dr. Elizabeth Cudney, industrial engineering professor at the University Missouri Science and Technology, we discovered and defined an observable paradigm that we had observed within other organizations that we identified as the "10-30-60 rule." We realized that there existed a commonality or repetitive phenomenon where approximately 10% of most enterprises are typically assigned to produce and deliver the customer value proposition, while about 30% of the enterprise were involved directly in supporting the value add, and the remaining 60% were in place to provide all the remaining service and support functions. Although these portions are not absolutes, they predictably hover around such proportions. Since discovery of the 10-30-60 rule in 2007, I've been able to observe and validate its existence through projects and activities with a number of different and diverse organizations.

When applying affordability to products, services, and systems, one should design and plan for solutions using the components of the affordability hierarchy and utilize what is known as the affordability architecture or "the house of affordability" (Figure 1.8).

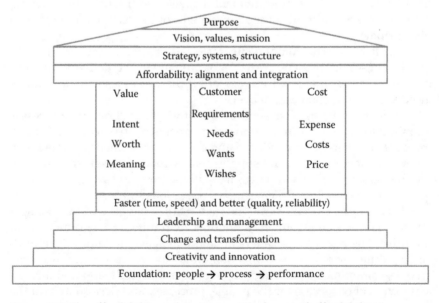

Figure 1.8 The affordability architecture or "the house of affordability."

This infrastructure illustrates, starting at the top, the strategic levels, down through the operational levels, to the tactical levels. It elucidates the architectural elements that must be in place to ensure and assure affordability.

Purpose: Each successful organization must have a clear reason for being. Also, the worth and value of the organization must be distinct, unique unto itself, and provide its intention. Communication of purpose must be clear and inherent for each function, and each individual must thoroughly understand the intent.

Vision, values, mission: The vision clarifies the direction and what the organization aspires. The values define how the organization operates, maneuvers, and trades within its business dealings and practices. Shared values include purpose and direction, as well as values and ethics. The mission expresses what must be done now and in the near-term or short-term future.

Strategy, systems, structure

Strategy: Expressed as a plan of action or policy designed to achieve a major aim. Affordability as a major aim is the foundation of a strategy for continuous improvement. In concert with Kaplan's and Norton's balanced scorecard, qualitative and quantitative metrics and measures can be established and monitored to achieve success:

- *Customer*—Whom we serve
- *Business*—The reason we do what we do
- *Process*—The work of the business and the work of providing value for the customer
- *People*—Those who provide the value and ensure our existence

Structure: Defined by the construct, configuration, and arrangement of the organization's resources.

- Organization is an organized body of people with a particular purpose, especially a business, society, association, and so on, that operate in an integrated and collaborative fashion to deliver value to a customer.
- Governance broadly refers to the mechanisms, processes, and relations by which organization is controlled and directed. Governance structures and principles identify the distribution of rights and responsibilities among different participants in the corporation (such as the board of directors, managers, shareholders, creditors, auditors, regulators, and other stakeholders) and include the rules and procedures for making decisions in corporate affairs. Corporate governance includes the processes through which corporations'

objectives are set and pursued in the context of the social, regulatory, and market environment. Governance mechanisms include monitoring the actions, policies, practices, and decisions of corporations, their agents, and affected stakeholders. Corporate governance practices are affected by attempts to align the interests of stakeholders.

- Financials and prosperity are the two dimensions that sustain the organization over time. Financials, including accounting, manage the money and wealth of the organization. Prosperity is the result of solid financial practices combined with the capital accrued from delivering products and/or services customers consume and pay for.

Systems: Consisting of all methods and means for receiving input for transformation into output and delivery of value to the customer.

- Customer value systems are the processes and procedures of the enterprise that deliver value to the customer. They provide that which the customer seeks to consume. The epitome of success is dependent on creating and delivering that which the customer is willing to pay for and purchase.
- Customer value support systems are the processes and procedures that support the value systems. They are absolutely necessary for success. They may be internal or external or both. Often confused with nonvalue add, without such support, the customer value systems would fail.
- Organization support systems exist in every organization. These systems support the organizations administration and operations. They also support the leadership, management, and governance of the organization. Although traditionally viewed as most critical departments, within affordability, using a customer centered focus, these areas are seen as value support functions.

In addition to strategy, structure, and systems, in the McKinsey 7 S model are skills, staff, and style, affectionately referred to as the "soft Ss" that also enhance the affordability infrastructure. Within the affordability context, these soft Ss embrace the people side of affordability.

- Culture and success are two tightly related dynamics within affordability. Although it may be a much overused cliché, a culture of "success breeds success," it's quite fitting within affordability. Successes or victories are something to celebrate and enjoy, losses are something to learn and grow from, and both must be cherished as they disseminate wisdom.

- Partnerships and relationships within affordability guarantee a continuity of service for customers and a continuous stream of flow for suppliers. Although customers are the focal point for care, delivery, and production, and suppliers are often neglected, both are critical from downstream and upstream perspectives. For products and services to flow smoothly through the enterprise, both ends of the spectrum are key critical and core significant. In fact, partnerships and relationships with all stakeholders throughout the enterprise are vital.
- Adaptability and flexibility are directly related to longevity. "Its not the strongest species that survives, or the most intelligent, but the most responsive to change" (Charles Darwin). It takes adaptability and flexibility to reach longevity. Those organizations with rigid and fixed paradigms often stagnate due to what Joel Barker calls "paradigm paralysis." The longest-lived organizations have had to adapt and flex for several hundred years or more.
- Community and environment are two areas that position an institution as a living organization. Investing resources and engaging in the local community and the surrounding environment pump a life blood into the society and surroundings.
- Learning and growth, although last on this list, is a key factor in motivating people and provides what Dr. Rosabeth Kanter calls the 3Ms: meaning, mastery, membership. It provides for individuals' enlightening and capability. The people who provide the value are, in another often overused cliché, truly "the most important asset."

Completing the "roof" of the house of affordability or strategic canopy, alignment and integration serve to incorporate all of the human, nonhuman, and systems dimensions directing and enabling the operational and tactical elements to operate. It connects value-added resources with value-added support resources with the organization support resources. It combines the purpose with the direction with the means. It communicates and conveys the "whom do we serve," with the "what to do," and the "how to do it" and designates the "who's going to get it done" even providing the "how do we know when we're successful?" Often when I get involved with organizations, two "problems" are cited as major issues: communication and division/separation of departments (terms used include "organization silos," "stovepipe organization," "islands of existence"). Communication is usually a symptom and the root cause often emerges as alignment clarity, and separation is often a sign of lack of integration and the absence of resource optimization.

The three pillars that sit upon the operational and tactical layers of the house and serve to support the strategic cover provide the basic elements and the primary focal points of affordability.

- The pillar of value provides intent, worth, and meaning for the organization. It is the reason that the "house" remains relevant. An organization that doesn't know its value is destined to become obsolete. Eastman Kodak, Blockbuster, AOL, Liggett Drugs, and Wang all lost their edge and suffered the consequences. A thriving institution and company inherently knows its value and provides that value to those who are more than willing to exchange wealth and capital for that value.
- The customer pillar, clearly in the middle, serving as the fulcrum or center for the balance of strategy, is the point at which the primary emphasis of affordability is placed. Bookended by value and cost, customer exists as the reason for ongoing existence. By meeting and exceeding the customer requirements, needs, wants, and wishes, continued existence is ensured.
- The pillar of cost helps to keep the organization relevant and viable. Often, with success comes abundance, excess, complacency, and slothful interdisciplinary fiduciary habits. Cost vigilance using conservative financial practices, and an ever-present attention to preserving wealth, reducing cost, and optimizing investment, will sustain treasure and insure permanence. Cost can be viewed internally as expense and cost to provide value, but also externally as customer cost and price. By reducing cost internally and externally, competitiveness can be assured.

Faster and better is synonymous with speed and quality. Speed can also be described with the terms velocity and responsiveness. In today's global economy, availability and accessibility of products and services can be a competitive advantage. In addition, "quality over the past 20–30 years has experienced a worldwide epidemic of sorts, and if you don't catch it, you may not survive" (Joel Barker quote).

Leadership is a core component for every organization. Today, worldwide, there seems to be a lack of leadership and an overabundance of management. Leadership is about people, while management is about things. In 1982, at a conference in which I was speaking, I heard the keynote speaker, Retired Rear Admiral, Grace Hopper say, "You manage things, and you lead people. People are unmanageable, they must be led!" Leadership sets direction, aligns resources, motivates people, communicates the message, and executes the plan.

Change and transformation is a leadership responsibility. This is pertinent to moving an organization forward. Although the intent of change is to advance the organization, care must be taken to mitigate fear that may arise. Transformation accomplishes required change, and the formulized components of transformation must all be in place: vision/mission/purpose, leadership, people, processes/resource, design/plan.

Organizations that are not adaptable and flexible resist change and transformation, and typically die. The most resilient organizations embrace change and transformation thoughtfully, methodically, and carefully with designed and planned rationale and strategy.

Creativity and innovation provides solutions to challenges, delivers answers to questions, and affords opportunities for ongoing success. Creativity is about generating the ideas addressing the challenges, questions, and opportunities. Innovation is how to implement those ideas to address and mitigate the challenges, questions, and opportunities. Although sequentially related, both are critical and stable systems for incorporating each one is important. First, through creativity, the idea must be generated, then innovation must take place to implement that idea and maintain the solution, resolution, or remedy.

The 5P Foundation: People, Process, and Performance are cornerstones and footing of affordability. People drive the process(es). The result of the process(es) is performance. The outcome of performance is profitability, both socially and financially. The social dimension of profitability is qualitative and people related. The financial dimension of profitability is quantitative and monetary related. The outcome of profitability is prosperity for the organization, the organization's partners and suppliers, and the people. The consequence of profitability is folded back to the people and the cycle continues. The fuel and energy that powers this economic engine is purpose (Figure 1.9).

It logically follows that, in order to design in affordability, one must begin with the customer. The needs, wants, wishes, and requirements of the customer must align and be in synch with both the value and cost of any organization. Speed and quality naturally relate to value and cost,

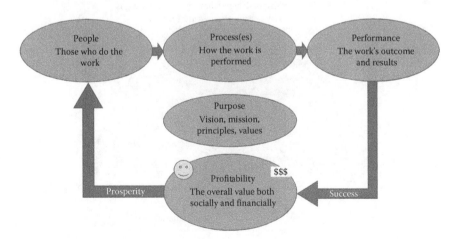

Figure 1.9 The 5Ps: a cyclical system and function of success.

and leadership must thereby tie them all together. To accomplish afford-ability, change and transformation may be necessary, and creativity and innovation are compulsory for solving problems and providing customer solutions. To complete the design, the resources required, the processes needed, and the performance criteria should all be clearly understood (Figure 1.10).

While implementing the design, the first step of the plan should clearly define and describe the people, processes, and performance of the enterprise and systems affected. Creativity and innovation should be invoked to spell out and articulate the change and transformation neces-sary. Leadership should be engaged to carry out and execute the plan. Time, quality, and cost should serve as quantitative parameters, while value, customer, and market needs should serve as qualitative factors. The result is a product or service delivered to the customer, produced and provided by the organization, that meets and exceeds the requirements (Figure 1.11).

To accomplish this endeavor, affordability provides a variety of tools, toolboxes, techniques, and methods. What follows is but a sampling of the implements and means, and how they are positioned in relation to the three-point aim (Figure 1.12).

In 2007, Dr. Elizabeth Cudney and I were invited by the Institute of Industrial Engineers to serve Northrop Grumman in its pursuit to implement Lean principles and practices. The program was champi-oned by Northrop Grumman Corporation's Senior Vice President George Vardoulakis and Program Lead Mr. Dave Armbruster. Although attempts at incorporating Lean had failed in 2000, 2002, and 2004, a conscious

Figure 1.10 Affordability design is customer driven.

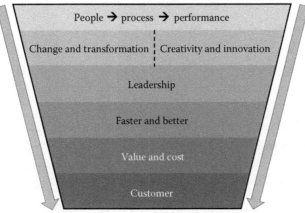

Figure 1.11 Affordability implementation is people → process → performance driven.

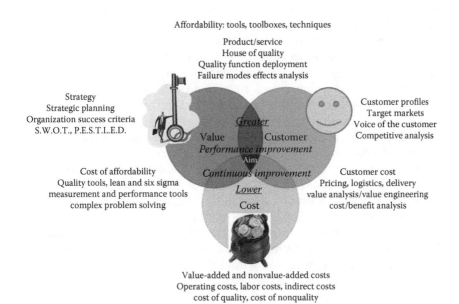

Figure 1.12 Affordability—tools, toolboxes and techniques.

Example: Northrop Grumman
60% of the overall statement of work for the
F/A-18 "Super Hornet"

Program: "affordability" (2007–2009)

· Metrics/measures	2006	2010
· Production capability	42	62
· Cycle time	1 per 5.5 days	1 per 4.0 days
· Quality	5σ	6σ
· Customer price	$55 M	$49.9 M
· Customer base	United States	United States, Australia, and others

· Employment: 1100+ in El Segundo, CA plus more than 700 parts suppliers

Note : Original scheduled plant closing 08/2010 →plant life extended to 2020.

Figure 1.13 Northrop Grumman F/A – 18 Super Hornet.

decision was made in 2006 to try one more time. What began as a Lean project, eventually evolved into the Program for Affordability. The results and outcomes speak for themselves. Figure 1.13 illustrates the before and after of the event that lit the fuse for realization and discovery of the theory of affordability.

The value was based on a U.S. Department of Defense (D.o.D.) need for a resilient, effective, competitive defense aircraft that served both navy and Marine Corps requirements. Throughout the project, the value was improved, the customer base increased, and the cost (internal and external) was reduced. In August of 2010, the U.S. government ordered 124 more planes. At that time, Assistant Secretary of Defense Dr. Ashton Carter distributed a memo to all D.o.D. acquisition personnel defining how to establish better buying power for the D.o.D., and cited the F/A—18 as an example of affordability (Figure 1.14).

The purpose of affordability is to set the aim and direction for continuous improvement through the lenses of three distinct components: value, customer, and cost. And, to remain vigilant and active to constantly increase value, while always seeking to exceed customer expectations, and to steadily reduce execution costs as well as relentlessly diminish customer costs. This is achieved through a constancy of incessantly improving the offerings and systems of the organization, the operations of the organization, the alignment of the products and services of the organization, and the systems of the organization with the value and value proposition, the target customers, and the target markets. The results: customer/market growth and loyalty, business profitability and expansion, process excellence and perfection, people motivation and devotion.

Affordability example: Northrop Grumman
F/A-18 "Super Hornet"

Dr. Ashton Carter
Memo to All D.o.D. Acquisition Personnel

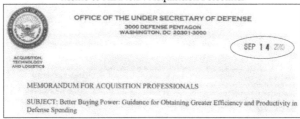

On Page 7, Par. 3

This was an affordability program result	The Navy, for example, recently concluded negotiations for a multi-year procurement of 124 F/A-18 strike fighter and E/A-18 electronic attack aircraft, which will yield over $600 million (greater than 10 percent) savings to the Department and the taxpayer. The F-18 program was able to drive down cost for each lot of aircraft procured in the framework of a fixed-price incentive contract that meets the Department's objectives for realistic costs, reasonable profit, a 50/50 shareline, and a 120 percent ceiling.	Value: over $600 MM savings to taxpayers

Figure 1.14 Better buying power. Dr. Ashton Carter memo.

For most, the tenets of affordability have always existed; however, affordability is a new comprehensive theory, idea, and concept for continuous improvement, incorporating and integrating value, customer, and cost for organization survival and longevity. Some will readily adopt this new philosophy, while some will wait and see if this is just another "strategy-du-jour," and yet others will resist this new idea as just another passing fad of performance improvement. The results and outcomes speak for themselves. The people who have experienced and realized affordability know it and comprehend its power. The challenge for any organization new to affordability is to create a constancy of purpose for improving products, services, and systems, to adopt this new philosophy, lead the organization through implementation, and institute a culture of excellence through change and transformation.

chapter two

Customers
Who really comes first?

> There is only one boss. The customer. And he can
> fire everybody in the company from the chairman
> on down, simply by spending his money some-
> where else.
>
> **—Sam Walton**

The customer is the center post of the house of affordability—the
focus, concentration, and emphasis of the foundation and framework.
Although value and cost are as important as the customer, and at the
same level of the core model, the customer always comes first in terms
of "keeping the lights on" and "making the money." Serving the customer is the primary goal, be it delivering a product or providing a service. The ultimate and ideal objective is to be the customers' only choice.
However, this is not a reality one should expect; hence, one should be
aware of the customers' requirements, needs, wants, and wishes, as
well as the competitors' offerings in detail. In the past, I've relied upon
various tools and techniques to discover the requirements and improve
product and service offerings. Customer requirements are needed for
defining functions, features, competitive advantages, and unique qualities in any product or service. Customer and value primarily determine
requirements (Figure 2.1).

Customer and cost determine price. The financial aspects complete
the equation of customer demand (customer demand is satisfied when
conformance to requirements are met at a price that is acceptable to the
customer. The result: customer buys). When the requirements are correct
and the price is right, the customer purchases the product. It is not only
what the customer is willing to pay for (i.e., the product or the service), but
also offered at a price the customer is willing to pay initially and over the
time the product or service is maintained. It does get more complicated
when the correct product or service is available from the competition at
the same price. When competition matches a provider's product or service
at the same price, other factors such as timing, availability, logistics, and
additional services included come into play. This is the reason a customer-
focused approach is mandatory for affordability.

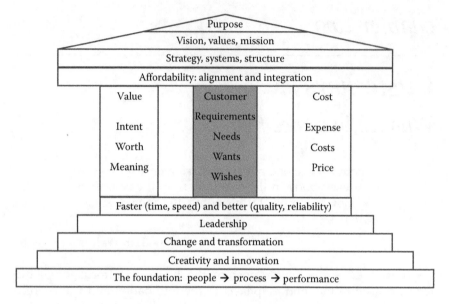

Figure 2.1 The affordability architecture or "the house of affordability."

Customer focus

Affordability defines customer focus as the center, aim, and target for the organization's unique value proposition. An organization that is fully customer focused designs its value proposition to meet and exceed the customers' expectations while operating with an attitude and culture of continuous improvement both internally and externally. For example, traditional manufacturing bases its production on the scheduling of process steps and material cost to push capacity and optimization, while contemporary manufacturing bases it processes' rhythm on demand, meeting the pace of customer consumption and replenishment of materials. Customer-focused, value-centered organizations operate very differently than internally focused, organization-centered establishments (Figure 2.2).

Organizations, businesses, companies, corporations, institutions, and establishments that operate using a customer-focused and value-centered design core are easy to recognize and enjoyable to work for, work with, and purchase products and services from as a customer.

- *Success is based on overall performance:* This includes metrics measuring customer, financial, process, and people performance. Qualitative measures for all types of people satisfaction; customer, employee, partners and suppliers. Quantitative measures for all types of

Customer focused vs. internally focused value centered vs. organization centered

Customer focused: value centered	Internally focused: organization centered
• Success is based on overall performance	• Success based primarily on financials
• Fulfill customer needs	• Fulfill organization needs
• Conformance to requirements	• Profit first
• Lower cost, reinvest	• More sales
• Compliance with standards	• Sales and marketing
• Customer provisioning	• Customer response
• Continuous improvement	• Continuous adjustment
• Operates offensively	• Operates defensively
• Long-term growth and success	• Short-term gains
• Enterprise and partner/supplier success	• Organization-only success

Figure 2.2 A comparison between a customer-focused, value-centered organization and an internally focused, organization-centered establishment.

quantitative performance are time/on time, quality, revenue/cost/ profitability. Additional measures are community, environment, competence, information, strategy, markets, measurement, creativity, innovation, competition, etc.

- *Fulfill customer needs*: Quality is seen as conformance to requirements, compliance with standards, and continuous improvement. Quality products and services are created and delivered based on the needs, wishes, wants, and desires of the customers.
- *Conformance to requirements*: When creating and providing products and services, attention to what is required by the customer is of the utmost importance to designers, engineers, producers, and providers. Requirements drive what is designed and produced to be delivered.
- *Lower cost, reinvest*: This is a philosophy of sound fiduciary responsibility. Reinvestment of funds recovered through cost reduction allows for more products and services robust and responsive to changing customer requirements.
- *Compliance with standards*: Stabilize, standardize, and sustain is a rally cry of affordability organizations. Consistency and constancy establish behaviors that often lead to permanence and strength. The antonym is instability and chaos.
- *Customer provisioning*: This behavior and habit anticipates customers' needs and provides product and service proactively. An example comes from Toyota during the early days of the release of the Toyota Prius. When first ordered, Toyota proactively managed the customers' expectations by giving a delivery date of 3 months lead time and delivered in half that time. Toyota realized it could meet a demand of customers preordering cars in 3 months; however, they were able to deliver in half that time.

- *Continuous improvement*: A culture of ongoing ever "Make Better!" organizations of this characteristic are never satisfied with status quo and seek to improve processes, products, services, people, and the entire enterprise.
- *Operates offensively*: The term offensively is not to infer rudely or abusively, but more along the lines of proactively and preemptive, anticipating the customers', markets', partners', and suppliers' needs. This is a positive and upbeat method for doing business.
- *Long-term growth and success*: Companies that have endured time have maintained this commitment and attitude. It goes without saying, such companies are proven to be effective in the long run and successful for the "long haul."
- *Enterprise and partner/supplier success*: Many companies serve their customers and serve themselves, but excellent companies of affordability also serve their enterprise as well as their partners and suppliers.

In addition to a customer-focused and value-centered culture, and as a result of my meeting and being exposed to the philosophy and teachings of Dr. W. Edwards Deming, I've created the 14 principles of affordability:

1. Purpose is the reason for existence.
2. Customers are the center and those whom we serve.
3. Products and/or services are the vehicles that sustain our existence.
4. People are the solution to challenges of our systems.
5. Leadership is responsible and accountable for the outcomes and results.
6. Creativity and innovation is the foundation of our future.
7. Facts are the basis of our decisions, truthful information is the basis of our knowledge.
8. Strategy provides the direction.
9. Systems are the means of success.
10. Structure clearly defines the playing field as well as the boundaries and constraints.
11. All work is a process where the quantitative dimensions of scope, time, quality, and cost are measurement components of success, and qualitative satisfaction measures of customer, people, and suppliers/partners are components of excellence.
12. Operations/tactics are achieved through the execution of plans developed strategically, operationally, and tactically that link everyone in the organization to the strategic plan and design.
13. Results are the indicators of achievement and accomplishment of our work.
14. Profitability is the outcome of performance in monetary, social, and cultural terms.

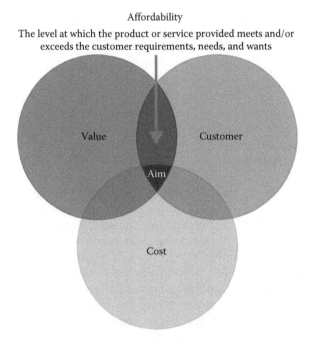

Affordability
The level at which the product or service provided meets and/or
exceeds the customer requirements, needs, and wants

Figure 2.3 The intersection of value and customer.

These principles dovetail nicely into the overlap of the customer and value pillars (see Figure 2.3). Customer requirements, needs, wants, wishes, and desires are the aim and target for fulfillment of the value proposition of the organization. Products and services aimed at customer markets should seek to meet and exceed all conditions and necessities. There are even times when close customer relationships can predict and anticipate emerging needs and upcoming shifts in market requisites. When value matches up to, and synchs with, customer realities, superior products and top-quality services are the likely result. In order to design and produce excellent products and services, the customer and market requirements need to be completely known and understood. There are several tools available to capture customer and value details of the requirements, needs, wants, wishes, and desires. Those that I've used the most are voice of-the customer (VOC), house of quality (HoQ), quality functional deployment (QFD), and failure modes effects analysis (FMEA).

Customer and value realization

In order to realize both customer and value goals and objectives, the affordability process of assess → design → implement → maintain should be used. A thorough assessment of the customer requirements, needs, wishes, wants, and desires must be executed. There are numerous tools available to perform this process. The one that I prefer is VoC.

Voice of the customer

There are a multitude of definitions of VoC, or sometimes referred to as VOC. Combining and integrating several of them, we have (some sources that have described it correctly include ASQ, iSixSigma, and even Wikipedia) the following:

- VOC is a term used in business to describe the in-depth process of capturing a customer's expectations, preferences, and aversions. Specifically, the VOC is a market research technique that produces a detailed set of customer wants and needs, organized into a hierarchical structure, and then prioritized in terms of relative importance and satisfaction with current alternatives. VOC studies typically consist of both qualitative and quantitative research steps. They are generally conducted at the start of any new product, process, or service design initiative in order to better understand the customer's wants and needs, and as the key input for new product definition, QFD, and the setting of detailed design specifications.
- Much has been written about this process, and there are many possible ways to gather the information—focus groups, individual interviews, contextual inquiry, ethnographic techniques, etc. But all involve a series of direct, structured in-depth interviews, which focus on the customers' experiences with current products or alternatives within the category under consideration. Needs statements are then extracted, organized into a more usable hierarchy, and then prioritized by the customers.
- It is critical that every organization that is part of the value stream and in direct support of the value stream be involved in the process. The designers, producers, and providers must be the ones who take the lead in defining the topic; designing the sample (i.e., the types of customers to include); generating the questions for the discussion guide, either conducting or observing and analyzing the interviews; and extracting and processing the needs statements.
- The VOC is comprised of actual customer descriptions in words for the functions and features that customers desire for products and services. In the strict definition, as relates to QFD, the term customer indicates the external customer of the supplying and delivering entity (i.e., the one delivering the products and/or services).
- The VOC is a process used to capture the requirements/feedback from the customer (internal or external) to provide the customers with

the best-in-class service/product quality. This process is all about being proactive and constantly innovative to capture the changing requirements of the customers with time.

- The VOC is the term used to describe the stated and unstated needs or requirements of the customer. The VOC can be captured in a variety of ways: direct discussion or interviews, surveys, focus groups, customer specifications, observation, warranty data, field reports, complaint logs, etc.
- These data are used to identify the quality attributes needed for a supplied component or material to incorporate in the process or product.
- The VOC is the feedback from your current and future customers indicating service offerings that satisfy, delight, and dissatisfy them. You can obtain the VOC through many different ways, including surveys, focus groups, interviews, listening posts, and mystery shopping.

VOC benefits

- Helps you focus on those items that are most valuable to your current customers
- Identifies gaps in the services you provide to the customer
- Allows you to focus on those items that will be most valuable to future customers
- Identifies potential improvement opportunities and the associated priorities from a customer viewpoint
- Allows you to decommission services that are of little or no value to your customers or future customers

VOC: How to ... execute the process

Step 1: Assess. Identify your customers, your future customers, and your potential. The customers you want to have in 5 years may be different from the customers you have today, especially in terms of numbers, growth, and prosperity.

Step 2: Design. Determine the tools and techniques you will use to gain feedback from your target customers. Surveys, focus groups, and interviews are popular tools used to gather the information.

Step 3: Implement. Incorporate the results of your data gathering and analysis into your service offerings, making continual improvements as appropriate.

Step 4: Maintain. Develop a schedule to reassess your customers on an ongoing basis to ensure your service offerings meet or exceed their requirements (Figure 2.4).

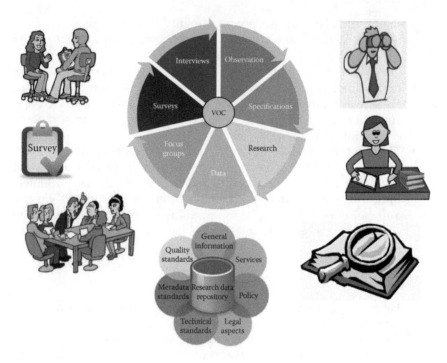

Figure 2.4 VOC assessment areas.

The house of quality

Brief: What is the HOQ?

The HOQ is a VOC analysis tool and a key component of the QFD technique. It starts with the voice of the customer. It is a tool to translate what the customer wants into products or services that meet the customer wants in terms of engineering design values by way of creating a relationship matrix (Figure 2.5).

- Typically, the first chart used in QFD.
- Data intensive and is capable of capturing large amounts of information.
- *Left side*: has the customer's needs.
- *Ceiling*: has the design features and technical requirements.
- *The roof*: a matrix describing the relationship between the design features. Used to show how the design requirements interact with each other.
- *Competitive section*: based primarily on the customer's perspective.
- *Lower level/foundation*: benchmarking and target values used to rank the "hows." These often contain the actions your organization will take to satisfy the customers.

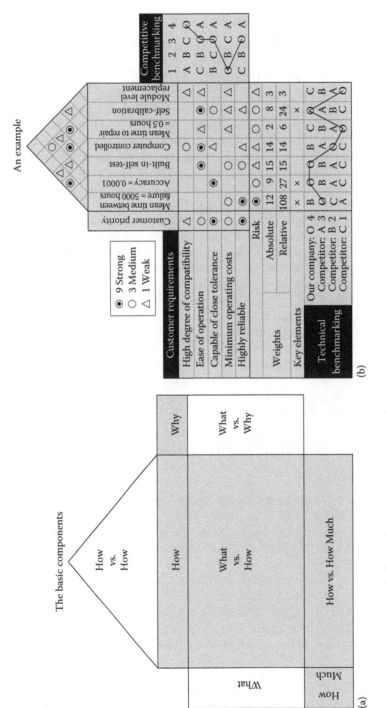

Figure 2.5 House of quality (a) The basic framework of the house of quality (b) House of quality example. (Excerpted from Jack B. ReVelle's Quality Essentials: A Reference Guide from A to Z, ASQ Quality Press, 2004, pp. 9–11.)

HOQ definition description

The HOQ is a matrix diagram resembling a house used for defining the relationship between customer desires and the firm/product capabilities. It is the front end of the QFD process (QFD is explained later), and it utilizes a planning matrix to relate what the customer wants to know how an organization will satisfy those customer wants. It looks like a house with a "correlation matrix" as its roof, it is the customer wants versus product features and functions, with how much as the foundation and a competitor evaluation "porch." It is based on "the belief that products should be designed to reflect customers' desires and tastes." It also increases cross-functional integration within organizations using it, especially between marketing, engineering, and manufacturing.

The basic structure is a table with "whats" as the labels on the left and "hows" across the top. The roof is a diagonal matrix of "hows vs. hows" and the body of the house is a matrix of "whats vs. hows," with a base of "how much" of the "hows." Both of these matrices are filled with indicators of whether the interaction of the specific item is a strong positive, a strong negative, or somewhere in between. Additional annexes on the right side and bottom hold the "whys" (market research, etc.) and the "how muches." Rankings based on the whys and the correlations can be used to calculate priorities for the hows.

HOQ analysis can also be cascaded, with "hows" from one level becoming the "whats" of a lower level; as this progresses, the decisions get closer to the engineering/manufacturing details; this gets us into QFD defined and described later.

Quality function deployment

What is QFD?

"Time was when a man could order a pair of shoes directly from the cobbler. By measuring the foot himself and personally handling all aspects of manufacturing, the cobbler could assure the customer would be satisfied," lamented Dr. Yoji Akao, one of the founders of QFD (Figure 2.6), in his private lectures.

QFD was developed to bring this personal interface to modern manufacturing and business, and eventually, services. In today's global economy, where the growing distance between producers and users is a concern, QFD links the needs of the customer with design, development, engineering, manufacturing, and service functions.

QFD is

- Understanding customer requirements
- Quality systems thinking + psychology + knowledge/epistemology

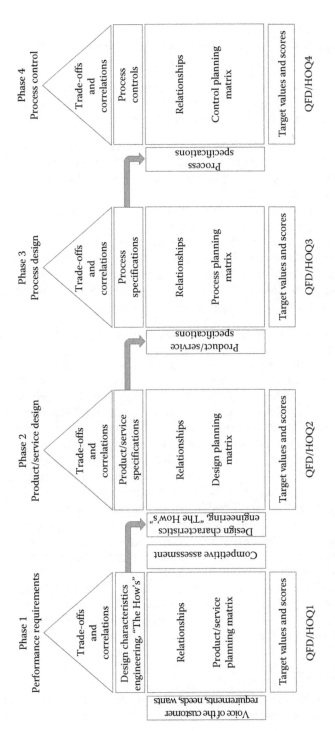

Figure 2.6 Quality functional deployment.

- Maximizing positive quality that adds value
- Comprehensive quality system for customer satisfaction
- Strategy to stay ahead of the game

As a quality system that implements elements of systems thinking with elements of psychology and epistemology (knowledge), QFD provides a system of comprehensive development process for

- Understanding "true" customer needs from the customer's perspective
- What "value" means to the customer, from the customer's perspective
- Understanding how customers or end users become interested, choose, and are satisfied
- Analyzing how do we know the needs of the customer
- Deciding what features to include
- Determining what level of performance to deliver
- Intelligently linking the needs of the customer with design, development, engineering, manufacturing, and service functions
- Intelligently linking design for Six Sigma (DFSS) with the front-end VOC analysis and the entire design system

There are many approaches to QFD, depending on the strategic purpose of the project.

QFD is a structured methodology and mathematical tool used to identify and quantify customers' requirements and translate them into key critical parameters. In Six Sigma, QFD helps you to prioritize actions to improve your process or product to meet customers' expectations.

QFD is a method to transform qualitative user demands into quantitative parameters, to deploy the functions forming quality, and to deploy methods for achieving the design quality into subsystems and component parts, and ultimately to specific elements of the manufacturing process, as described by Dr. Yoji Akao, who originally developed QFD in Japan in 1966, when the author combined his work in quality assurance and quality control points with function deployment used in value engineering (VE).

QFD is designed to help planners focus on characteristics of a new or existing product or service from the viewpoints of market segments, company, or technology development needs. The technique yields charts and matrices.

QFD helps transform customer needs (the VOC) into engineering characteristics (and appropriate test methods) for a product or service, prioritizing each product or service characteristic while simultaneously setting development targets for product or service.

Failure modes effect analysis

FMEA—also "failure modes" (plural)—in many publications was one of the first systematic techniques for failure analysis. It was developed by reliability engineers in the late 1950s to study problems that might arise from malfunctions of military systems. An FMEA is often the first step of a system reliability study. It involves reviewing as many components, assemblies, and subsystems as possible to identify failure modes, and their causes and effects. For each component, the failure modes and their resulting effects on the rest of the system are recorded in a specific FMEA worksheet. There are numerous variations of such worksheets. An FMEA can be a qualitative analysis, but may be put on a quantitative basis when mathematical failure rate models are combined with a statistical failure mode ratio database (Figure 2.7).

A few different types of FMEA analyses exist, such as

- Functional
- Design
- Process

Sometimes, FMEA is extended to FMECA to indicate that criticality analysis is performed too.

FMEA is an inductive reasoning (forward logic) single point of failure analysis and is a core task in reliability engineering, safety engineering, and quality engineering. Quality engineering is specifically concerned with the "process" (typically, manufacturing, production, assembly, etc.).

A successful FMEA activity helps to identify potential failure modes based on experience with similar products and processes—or based on common physics of failure logic. It is widely used in development and manufacturing industries in various phases of the product life cycle. Effects analysis refers to studying the consequences of those failures on different system levels.

Functional analyses are needed as an input to determine correct failure modes, at all system levels, both for functional FMEA or piece-part (hardware) FMEA. An FMEA is used to structure mitigation for risk reduction based on either failure (mode) effect severity reduction or on lowering the probability of failure or both. The FMEA is in principle a full inductive (forward logic) analysis; however, the failure probability can only be estimated or reduced by understanding the failure mechanism. Ideally, this probability shall be lowered to "impossible to occur" by eliminating the (root) causes. It is therefore important to include in the FMEA an appropriate depth of information on the causes of failure (deductive analysis).

FMEA is a systematic, proactive method for evaluating a process to identify where and how it might fail and to assess the relative impact of different failures in order to identify the parts of the process that are most in need of change. FMEA includes review of the following:

- Steps in the process
- Failure modes (what could go wrong?)

FAILURE MODE AND EFFECTS ANALYSIS

Item:	Product, Process, Procedure, etc.	Responsibility:
Description:	Title/Name	Prepared by:
Core Team:	Names on the Team Members performing the FMEA	

Name of Process Owner	FMEA number:	Page:	Tracking Number / Page #
Name of Preparer		FMEA Date (Orig):	Date
		Rev:	Revision

Process or Function Description	Potential Failure Mode	Potential Effect(s) and Severity of the Failure	Sev	Probability of the Failure to occur or likelihood the Failure will occur	Occ	The ability to detect the failure and mitigate	Det	RPN	Recommended Action(s)	Responsibility and Target Completion Date	Actions Taken	Sev	Occ	Det	RPN
The specifics of the function or procedure	Description of the failure	A definition of the effects created by the failure and its severity	1 to 10	The frequency of occurrence of the failure.	1 to 10	The level of capability to detect the failure within the process.	1 to 10	0							0
Examples	**Failures**	**Severity**		**Occurrence**		**Detection**		**RPN**	**Recommendations**	**Who and When**	**Actions**	**S**	**O**	**D**	**RPN**
Proc. 1	Desc. 1	A very severe failure	10	This failure occurs very frequently	10	We can not detect and trap this failure	10	1000	Apply a competent SWAT team immediately	Sue B. to Lead the team, Initial Action within 30 days.	The initial action was to detect and trap the failure	10	10	1	100
Proc. 2	Desc. 2	A moderate level failure	5	It occurs 50% of the time	5	Half the time we can detect it	5	125	Focus engineering and manufacturing on improving the process	John D. and Bob S., completion target mm/dd/yy	The first step of action was to reduce the occurrence and trap the failure	5	2	2	20
Proc. 3	Desc. 3	A minor failure	2	It occurs very little	2	We detect most of the failures	2	8	No Action Recommended	No Action Needed	No Action Taken	2	2	2	8
Proc. 4	Desc. 4	A somewhat moderate failure	4	It occurs somewhat frequently	4	We do not have a good method of detection	7	224	Fix the Process	Process Owner: Beth C., mm/dd/yy	Improve audit and validation procedures	4	3	3	36
Proc. 5	Desc. 5	A semi-severe failure	8	It does not occur often	8	We can easily detect it.	3	48	No Action Recommended	No Action Needed	No Action Taken	8	3	2	48
Proc. 6	Desc. 6	A moderate level failure	6	It occurs 70% of the time	7	We can not detect it about 3/4ths of the time	7	294	Correct the process and eliminate the failure	Billy B., mm/dd/yy	Re-Design and Re-Engineer the Process	1	1	1	1
Proc. 7	Desc. 7	This is a very minor failure	1	This failure is very rare	1	We can detect and trap it every time	1	1	No Action Recommended	No Action Needed	No Action Taken	1	1	1	1
::::::::	::::::::							0							0
Proc. N	Desc. N							0							0

Action Results

Figure 2.7 FMEA example.

- Failure causes (why would the failure happen?)
- Failure effects (what would be the consequences of each failure?)

Teams use FMEA to evaluate processes for possible failures and to prevent them by correcting the processes proactively rather than reacting to adverse events after failures have occurred. This emphasis on prevention may reduce risk of harm to both patients and staff. FMEA is particularly useful in evaluating a new process prior to implementation and in assessing the impact of a proposed change to an existing process.

Healthcare FMEA: FMEA was developed outside of healthcare and is now being used in healthcare to assess risk of failure and harm in processes and to identify the most important areas for process improvements. FMEA has been used by hundreds of hospitals in a variety of Institute for Healthcare Improvement programs, including idealized design of medication systems, patient safety collaboratives, and patient safety summit.

Problems and defects are expensive. Customers understandably place high expectations on manufacturers and service providers to deliver quality and reliability.

Often, faults in products and services are detected through extensive testing and predictive modeling in the later stages of development. However, finding a problem at this point in the cycle can add significant cost and delays to schedules. The challenge is to design in quality and reliability at the beginning of the process and ensure that defects never arise in the first place. One way that Lean Six Sigma practitioners can achieve this is to use FMEA, a tool for identifying potential problems and their impact.

FMEA: The basics

FMEA is a qualitative and systematic tool, usually created within a spreadsheet, to help practitioners anticipate what might go wrong with a product or process. In addition to identifying how a product or process might fail and the effects of that failure, FMEA also helps find the possible causes of failures and the likelihood of failures being detected before occurrence.

Used across many industries, FMEA is one of the best ways of analyzing potential reliability problems early in the development cycle, making it easier for manufacturers to take quick action and mitigate failure. The ability to anticipate issues early allows practitioners to design out failures and design in reliable, safe, and customer-pleasing features.

Finding failure modes

One of the first steps to take when completing an FMEA is to determine the participants. The right people with the right experience, such as process owners and designers, should be involved in order to catch potential failure modes. Practitioners also should consider inviting customers and suppliers to gather alternative viewpoints.

Once the participants are together, the brainstorming can begin. When completing an FMEA, it is important to remember Murphy's law: "Anything that can go wrong, will go wrong." Participants need to identify all the components, systems, processes, and functions that could potentially fail to meet the required level of quality or reliability. The team should be able to describe not only the effects of the failure but also the possible causes.

Criteria for analysis

An FMEA uses three criteria to assess a problem: (1) severity: the severity of the effect on the customer; (2) frequency: how frequently the problem is likely to occur; and (3) detection: how easily the problem can be detected. Participants or teams must set and agree on a ranking between 1 and 10 (1 = low, 10 = high) for the severity, occurrence, and detection levels for each of the failure modes. Although FMEA is a qualitative process, it is important to use data (if available) to qualify the decisions the team makes regarding these ratings. After ranking the severity, occurrence, and detection levels for each failure mode, the team will be able to calculate a risk priority number (RPN). The formula for the RPN is

$$RPN = Severity \times Occurrence \times Detection$$

Setting priorities

Once all the failure modes have been assessed, the team should adjust the FMEA to list failures in descending RPN order. This highlights the areas where corrective actions can be focused. If resources are limited, practitioners must set priorities on the biggest problems first.

There is no definitive RPN threshold to decide which areas should receive the most attention; this depends on many factors, including industry standards, legal or safety requirements, and quality control. However, a starting point for prioritization is to apply the Pareto rule: typically, 80% of issues are caused by 20% of the potential problems. As a rule of thumb, teams can focus their attention initially on the failures with the top 20% of the highest RPN scores.

Making corrective actions

When the priorities have been agreed upon, one of the team's last steps is to generate appropriate corrective actions for reducing the occurrence of failure modes, or at least for improving their detection. The FMEA leader should assign responsibility for these actions and set target completion dates.

Once corrective actions have been completed, the team should meet again to reassess and rescore the severity, probability of occurrence, and likelihood of detection for the top failure modes. This will enable them to determine the effectiveness of the corrective actions taken. These assessments may be helpful in case the team decides that it needs to enact new corrective actions.

The FMEA is a valuable tool that can be used to realize a number of benefits, including improved reliability of products and services, prevention of costly late design changes, and increased customer satisfaction.

Customer and cost

Complimentary to customer and value, customer and cost of the affordability model provides the typically "unheard of" opportunity for price reduction. One example is the increase of benefits without having to increase price. Adding on benefits offered at the same price is a cost/benefit improvement. Under affordability, efforts should be made to lower price through reduction of cost and increase in features, functions, and benefits. This approach ensures loyalty (Figure 2.8).

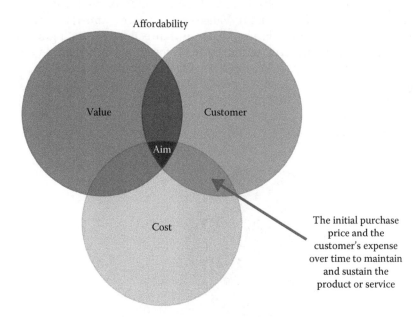

Figure 2.8 The intersection of customer and cost.

Value analysis/value engineering

Other than quality methods, Lean, and Six Sigma for reducing customer cost, there is another toolset I favor called value analysis (VA)/VE. When VA/VE is accompanied by product analysis and service analysis, customer cost reduction can be realized. VE is a systematic method to improve the "value" of goods or products and services using an examination of materials, variable overhead, labor, and related functions. Value, as defined, is the ratio of function to cost and all costs associated with the functions. Value can therefore be increased by either improving the function or reducing the cost. It is a primary tenet of VE that basic functions be preserved and not be reduced as a consequence of pursuing value improvements. Most value improvements are found in materials (see Figure 2.9).

The reasoning behind VE is as follows: if marketers expect a product to become practically or stylistically obsolete within a specific length of time, they can design it to only last for that specific lifetime. The products could be built with higher-grade components, but with VE, they are not because this would impose an unnecessary cost on the manufacturer, and to a limited extent also an increased cost on the purchaser. VE will reduce these costs. A company will typically use the least expensive components that satisfy the product's lifetime projections.

Due to the very short life spans, however, which is often a result of this "VE technique," planned obsolescence has become associated with product deterioration and inferior quality. Vance Packard once claimed this practice gave engineering as a whole a bad name as it directed creative engineering energies toward short-term market ends. To counterbalance such an effect, affordability supplements VA/VE with complete product analysis and service analysis that breaks down every facet to its salient components and uses the results for additional team analysis and diagnostics.

When used in a comprehensive way, all of the customer tools (i.e., VOC, HoQ, QFD, FMEA, VA/VA, and product/service analysis) ensure

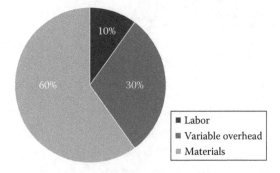

Figure 2.9 General cost breakdown: products.

consistency, stability, and reliability of products and services. Often, these affordability methods and techniques bring customers and providers closer together, sometimes even to the extent and level of customer as partner.

Customer as partner

Suffice it to say, not all customer relationships from a requirements and value perspective turn out fruitful and copacetic. I have had a few activities in the past where the customer suffered when needs, wants, and requirements were not well articulated, nor well known, nor understood, and out of synch with value and the value proposition. On occasion, other than my own customers and clients, I am asked to partner with other organizations to deliver services. Most of my lessons learned about customer requirements and impact of misunderstood needs, wants, and wishes came from organizations attempting to deliver products and services without fully knowing and understanding what the customer wanted and how those needs should be met and how the organization should be configured to deliver the services to the customer. I'd like to explain in more detail using specific case examples:

- A year or so ago, I was approached to serve as a Lean consultant to identify areas of waste and cut expense. My customer, who had a consulting firm, approached me to address one of his clients who happens to be one of the largest hospital systems in the United States. After a few days of being onsite with the customer, it became clear to me that what he wanted was to identify specific individuals who could be marked and "branded" for downsizing. The customer wanted to reduce cost. He wanted the cost reduction to take place in the next few months or as soon as possible. There was definitely a conflict between my partner/client and his customer. It brought back my old memories in corporate America where I acquired my distaste for layoffs and downsizings resulting from poor leadership and management. Although it was commonplace, if not scheduled annually, it was burned in my memory as an example of what not to do. Although I told him Lean does not mean less employees are needed, and there is no silver bullet, instant pudding, magic wand, nor free lunch, his requirements and my value proposition were not in synch. Lesson learned: true customer requirements can sometimes be different than reality, especially when you are partnered as a third party and in between a client and a customer who have different views of what is needed.
- In another case, and a most recent example, I was approached by a medium-sized consulting firm in Atlanta area to join them as a partner and serve a well-known healthcare system in the vicinity.

The customer requirements were quite clear and reasonable. In fact, the leader requesting the service had come from another healthcare system in the area that had gone through a similar transformation. It soon became apparent to me that the internal silos of disparate departments within the consulting firm, a firm that also specializes in organizational effectiveness, seemed to continually create barriers and roadblocks to our success as partners. Various incidents continued to fall across our path and get in the way of delivering customer service. First of all, the HR division requested that I provide 20 years of corporate tax records for proof of being an independent business consultant since I had been in that capacity since 1995. Then, the background checking and verification arm, a third party, could not get my information correct to check and verify my master's degree at the University of South Carolina in 1982. And, as time went on, with internal hurdles and obstructions, the client waited, and waited, and waited. Lesson learned: the customer will suffer when competing internal requirements get in the way of delivering service.

- As a final example of situations that can negatively impact the customer, I have an international incident instance that turned out to be quite insightful. I had been providing teaching, coaching, and mentoring services in the areas of leadership, strategy, and continuous improvement to organizations in the Middle East. Mostly Saudi Arabia and Qatar in the business sectors of banking, healthcare, education, and manufacturing. With the increased threat of terrorism and the actions of terrorists in the United States, Europe, and the Middle East, I requested that I receive direct roundtrip flights from Atlanta to Dubai on Delta Airlines in order to avoid interim stops in Europe or anywhere else. Several trips were on the agenda destined for: Dammam, Riyadh, Jeddah, Medina, and Doha, and the flights to these cities from Dubai were not long and scheduled frequently. In order to do business in Saudi Arabia, I had to have a sponsor and be part of an organization approved by the Saudi Arabian government. That organization made arrangements for all the logistics and support for each trip. As the time frame of the next trip approached, the organization seemed to have a great deal of difficulty booking the flights, even after I gave them the dates, times, and flight numbers of the Delta flights available from Atlanta to Dubai. I was told that those flights were completely booked even after I had verified there were at least 20–30 seats remaining on each one. I found out that behind the scenes the organization was being driven to use very cheap flights that required multiple stops and much longer time frames. Again, the customers suffered and the services couldn't be delivered. Lesson learned: partner with organizations designed for affordability!

Thank goodness, I have many more good case examples to share than the few bad ones I am able to offer. For the customer case examples (note: these are more strategic and operational than tactical) given later, I selected a variety of situations with different conditions and situations.

- While working for NCR, in the retail systems division during the early 1990s, I experienced a most memorable event that directly relates to customer requirements and customer partnership. Limited Brands, headquartered in Columbus, Ohio, used NCR products in their retail stores. As it is with almost all retail stores today, scanners and bar-coded tags have improved transaction performance, data reliability, and financial accuracy. One of their brands, Victoria's Secret, presented a new requirement to NCR for use in each of their many stores (there are now around 1100 stores worldwide). For convenience, ergonomic, and energy-saving reasons, they wanted a scanner that would operate dormant (to save energy) and wake up and scan a bar-coded tag when a store associate presented it to the scanner. They didn't want a bulky tabletop scanner, they didn't want a scanner gun requiring multiple trigger pulls, but they did want a portable scanner that was not fixed and permanent, that would also awake and scan, and go back to sleep when not needed. The scanner engineers came up with the 7890 Presentation Scanner that used a motion-activated sensor that would scan a bar code when it was placed in front of the scanner. It was designed to be ergonomically easy to handle, as well as small and portable, with a counter top cradle for ease of access and storage when not in use. Today, more than 20 years later, 7893 scanners (four generations later) can be purchased for use. Lesson learned: VOC, listening to the customer, partnering with the customer, and implementing the customer's wants, wishes, needs, and desires can pay off.
- Before the John H. Harland Company was merged with Clarke American Corp. to form Harland Clarke, I was engaged with them on a project focused on reducing defects encountered in the process of printing check for customers that were new to their client banks. This was a "sore spot" of dissatisfaction between Harland and their bank customers. The process was fraught with failure points, inconsistencies, and variation. The condition occurred when a new bank customer was issued a checking account by the bank, and quite frequently, the checks would arrive in the customer's mailbox with errors that required the customer to return the checks for reprinting. The result was a disgruntled bank customer and a discontented Harland bank customer. Since check printing and production was a major value stream for Harland, this opportunity loomed as their greatest headache and dissatisfier with this

type of customer. Our first step was to value stream map the process from the point of occurrence (the customer) upstream all the way through printing and design to the new customer check application process in the bank. Once the process was validated by all the people involved, we process mapped the entire process and performed a comprehensive process FMEA. We prioritized the failure modes by RPN, then began an extensive problem-solving effort to fix root-cause effects that created defects. Solutions included standardization of procedures, learning for all clients, error proofing of input tools (forms, systems, templates), in-process verification techniques, and data validation/verification methods. The outcome of the effort eliminated almost all possible failure modes and effects ... with the minor exception of rare human error conditions. Lesson learned: even when a procedure or process step is out of complete control of the organization, customer affordability tools can be applied to reduce defects, improve quality, and increase performance.

- I have found that, in my experience, the thorough use and utilization of the HOQ and QFD is uncommon. In the 2005 time frame, I was involved in a project with a company named FiberVisions. They create several different types of synthetic fiber for use in industrial and commercial applications. They produce the fibers that go into diapers for absorption, softness, and biodegradability. They produce the fibers that are placed in a patch under packaged chicken to absorb the liquids. They also produce the fibers that are woven into sanitary wipes, and in addition, many other applications. As a result of some DFSS training I provided, the leader of the effort, Aaron Gillory, decided to implement HOQ and QFD for responding to customer requests for new products. The top 10 customers were chosen, which represented 80% of the business and only 20% of the customer base. HOQs were created and QFD was applied. The consequence was impressive. What used to take 2–3 weeks or more to engineer and prototype a fiber ended up being only 1 day. Lesson learned: faster, better, more responsive to customer needs.

- Personally, I have always been, and will always be, a fan of partnering with the customer. It opens up the relationship for more trust, integrity, respect, and mutual success. My all-time favorite customer partner is the Institute of Industrial Engineers (aka the IIE and now having a subsidiary named IIE Solutions or IIES, which will both have name adjustments soon to IISE and IISES, respectively). My education and instructional services are provided to the IIE for public and private courses, and my consulting services are delivered by way of IIES projects. Since I partnered with them more than 10 years ago, we have developed strong Lean and Lean Six Sigma curriculums and achieved compelling results from project outcomes. It is the type of relationship where both parties prosper.

Once gained, a customer should be maintained, sustained, and served. I have found that the old adage is absolutely true: "It is ten times more difficult to get a new customer than to keep an existing one." To accomplish this, I try to always use the "customer golden rule": do not only treat your customers the way that you like to be treated but also treat your customers the way they would like to be treated, and better.

Reference

ReVelle, J. B. (2004) *Quality Essentials: A Reference Guide from A to Z*, ASQ Quality Press, Milwaukee, WI, pp. 9–11.

chapter three

Value and cost
The role of purpose and worth in affordability

> Try not to become a man of success, but rather try to become a man of value.

> **—Albert Einstein**

The attributes and synonyms for value include worth, importance, significance, usefulness, meaning, merit, benefit, appeal, attraction, and pull. Other attributes of value are price, cost, charge, rate, and profit. When studying cost, you will find some of the same alternatives: price, charge, rate, fee, amount, and expense. So, to me, it's not so surprising that value is often confused with cost and frequently used in the same context and intent. In affordability, value is most closely related to worth and benefit, while cost is expressed through more monetary characteristics of price and expense. Affordability, although centered on customer, is framed by value and cost, and the two supporting pillars provide that balance for integration for continuous improvement. Value contains cost as expense, and customer contains cost as price, so as cost contains both customer and value customer contains value as requirements. That's how affordability functions when implemented correctly (Figure 3.1).

Traditionally, and as an outcome of the twentieth century, the concepts of value, price, expense, and cost became very blurred and convoluted. And often, during the first part of this twenty-first century, we seem to be using the terms interchangeably from an internal perspective, especially value and cost. Most recently, I was facilitating a group of Lean Green Belts, and we were engaged in a conversation around value that started out as an amiable discussion, which in turn switched to a debate, and subsequently morphed into a disagreement, then finally shifted to a point of confusion about the use of the term "value" and the term "cost." I was introducing the concept of "value" to the group, and every time I used the term value, a couple of members of the group instantly translated it into their own understanding as cost. So, when I introduced the concepts of value-added resources, value-added support resources, and nonvalue-added resources and suggested we should consider increasing

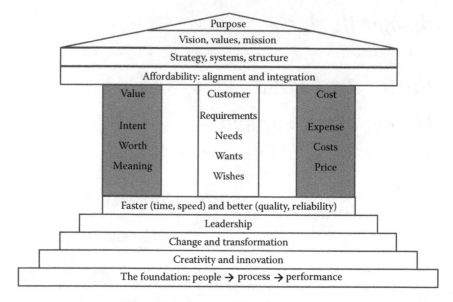

Figure 3.1 The affordability architecture or "the house of affordability."

the value-added areas, optimizing the value-added support areas, and seeking to eliminate the nonvalue-added areas, the two "value/cost participants" vehemently objected. They quickly made a connection in their own mind from history and experience: value → cost → employee downsizing and layoffs. Although 90% of the group understood the intent and implication of the proposal, the 10% that quickly rejected the concept clearly displayed their archaic value-as-cost paradigm, and value as defined in terms of cost alone. Such an out-of-date view tends to lead organizations down the path of failure due to focus on financial factors alone, not considering the entire spectrum and gamut of value.

There are instances where value can be provided at no price nor cost. Compliments and heartfelt gratitude can be offered by leaders that in turn stimulate motivation at no price nor any cost. Intent can be expressed and demonstrated through the actions and behaviors of all members of an organization at no price nor cost. Worth can be conveyed and communicated without having to spend money on elaborate branding campaigns or lavish marketing promotions. Meaning can be recognized by partners, colleagues, and associates over time. Value, characterized by intent, worth, and meaning, has a higher significance than price or cost. However, it should not go unsaid that value is balanced in the marketplace by price, cost, and expense, and is determined by the customer and the market. However, value can be defined, determined, and developed within an organization as a competitive advantage and a criterion for success and growth.

Cost, although related to price, may also have a level of expense associated with it. A certain level of value may come at a particular price, cost, and expense. A product or service containing such value will have to be sold at a particular price that includes some level of cost and expense. A wise man once told me, free always comes at a cost ... and if it's absolutely free, it has no realized value. In order to provide a product or deliver a service, an organization must utilize resources and materials that come at a cost and expense. To cover for that cost and expense, a particular price must be collected or the organization will operate at a loss, and eventually fail.

The area where value and cost overlap is where expense and the price to pay for delivering value exist. Figure 3.2 illustrates where the pillars of value and cost come together. Optimally, only value-added costs and value-added support costs would be accrued in this zone. But, all too often, nonvalue costs appear that negatively impact affordability and usurp funds from value-added functions.

The origins of this dimension of affordability can be traced back to Henry Ford. His attention to time, quality, and cost set the groundwork for affordability. Later in the century, Shigeo Shingo said, "There are four purposes of improvement: easier, better, faster, cheaper." It became more clearly evident during WWII when there were shortages of skilled labor, raw materials, and component parts that were common realities

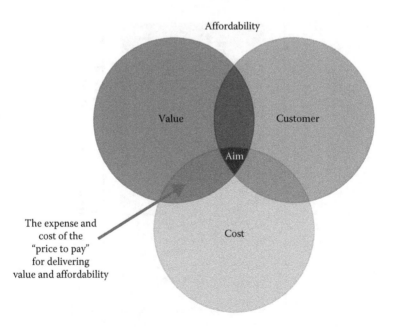

Figure 3.2 The intersection of value and cost is expense.

for production organizations. Lawrence Miles, Jerry Leftow, and Harry Erlicher at General Electric observed that acceptable substitutes often reduced costs, improved the product, or both. What started out from a necessity turned into a systematic process that led to VE and VA. When left to stand alone, this approach and technique is often criticized due to its narrow scope. However, when both are simultaneously considered with the customer pillar in affordability, a balance with requirements and price results in a formidable application method. Solely focusing on the overlap of value and cost, we discover a point where we realize that the expense necessary for delivering the value to meet customer requirements and the customer price required to achieve profitability may not coincide. Although this factor is not VE/VA in the purest sense, affordability uses the VE/VA philosophy and technique for understanding the current state and define the future state for improvement, with a plan to achieve the desired state, from a pure value/cost viewpoint. First and foremost, the value-added costs, value-added support costs, and nonvalue costs should be identified, understood, and categorized as either affordability costs or nonaffordability costs. The model in Figure 3.3 illustrates where affordability and nonaffordability can be encapsulated, mapped, and understood.

Keep in mind, since there are also cases and conditions where there is mandatory cost that provides no value, we must include that as an aspect that can be changed and improved to realize greater profitability.

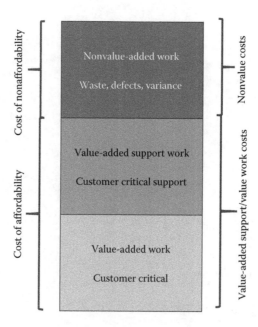

Figure 3.3 Value and cost of affordability.

In addition, when processes operate with wide variation and produce defects, there definitely exists a mandatory cost required to stay in business while the process is producing defects and rework is required. This is the cost that comes with nonaffordability. This particular cost has been described by Philip Crosby as the "cost of quality" (discussed in a later chapter) and is summarized by illustrating cost in terms of quality and work. For affordability, value and cost can be similarly illustrated in terms of affordability and nonaffordability, and value-added/value-added support costs and nonvalue-added costs.

The aim in this condition is to reduce the cost of nonaffordability, while increasing the spend on affordability. Strategically, an improvement project should focus on increasing value-added work and optimizing value-added support work, while reducing nonvalue-added work. The target condition, or desired future state, is a greater portion of the funds being applied to value-added work, with funds being better applied or optimized for value-added support work, while the expense of nonvalue-added work is minimized, leaving funds available for reinvestment or any other discretionary use. Figure 3.4 illustrates the change and transformation consequence when value and cost are expressed in terms of an affordability project.

As discussed in Chapter 2, value and customer integrate at the point of requirements, needs, wants, and wishes. To deliver that value and exceed customer expectations, there are associated costs with both elements. The expense and cost, or the "price to pay" to deliver value, is where value and cost intersect. The purchase price and the cost to maintain what a customer pays for a product or service is where customer and cost intersect. The ultimate goal is to increase the value proposition while reducing the value costs and improve the customer offerings while reducing customer costs. Programs and initiatives of affordability containing numerous projects yield an ever-increasing fund for more discretionary investment. Within the context of affordability, we see value and cost related to improvement and increase of profitability (Figure 3.5).

This leads us down the path of the old methods and beliefs versus the new approaches and techniques. In the 1980s and early 1990s, I was taught, and believed, the primary method for increasing profit was to increase sales, which provided growth. From about the mid-1990s until today, I began to realize that there is another, more efficient way to increase profit, without increasing sales. As a result and outcome of numerous projects and programs, I was able to see an improved approach for reducing cost, while increasing profitability, and growing demand, and improving market share, while yielding annuity investment monies for discretionary spending in areas of targeted performance improvement. An industry colleague of mine, Dr. Dan L. Shunk, has a model that clearly articulates this new way of thinking and acting.

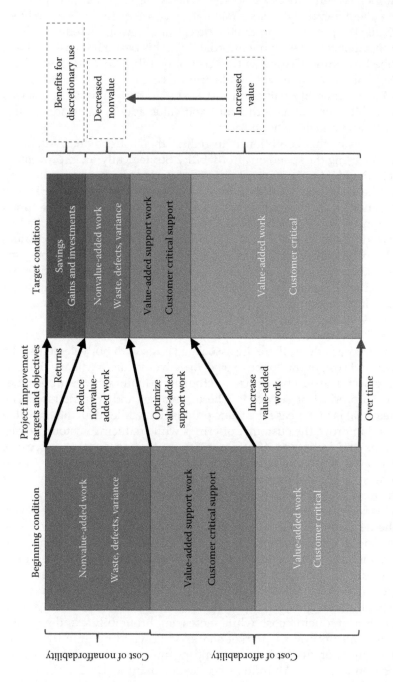

Figure 3.4 The relation of value and cost of an affordability project.

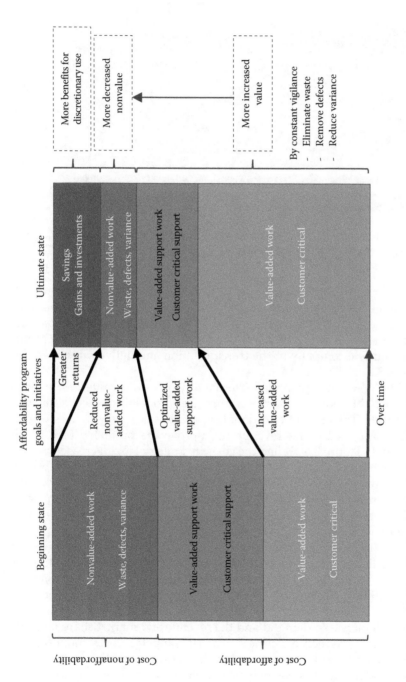

Figure 3.5 The relation of value and cost of an affordability programs.

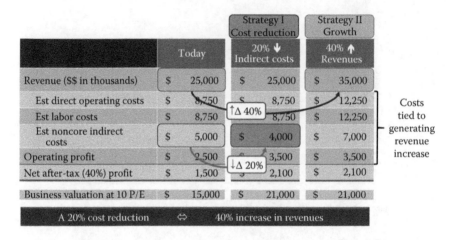

	Today	Strategy I Cost reduction 20% ↓ Indirect costs	Strategy II Growth 40% ↑ Revenues	
Revenue ($$ in thousands)	$ 25,000	$ 25,000	$ 35,000	
Est direct operating costs	$ 8,750	$ 8,750	$ 12,250	Costs
Est labor costs	$ 8,750	$ 8,750	$ 12,250	tied to
Est noncore indirect costs	$ 5,000	$ 4,000	$ 7,000	generating revenue
Operating profit	$ 2,500	$ 3,500	$ 3,500	increase
Net after-tax (40%) profit	$ 1,500	$ 2,100	$ 2,100	
Business valuation at 10 P/E	$ 15,000	$ 21,000	$ 21,000	

↑Δ 40% ↓Δ 20%

A 20% cost reduction ⇔ 40% increase in revenues

Figure 3.6 Value/cost strategies: Strategy 1: Cost reduction (New) and Strategy II: sales growth (Old). (Courtesy of Dr. Dan L. Shunk, ASU.)

In fact, when I asked him for the right to use his model, he told me, "We should be telling the world what we do and how we do it." So, following his suggestion, I'm using his model to tell the world what we're doing and how we're doing it. The model was developed using a real-time case example company in the transportation and delivery business (see Figure 3.6).

The new strategy, under affordability, uses the value/cost expense approach of reducing costs to increase profitability (see strategy I, Figure 3.6). The old and traditional approach, sales growth, uses primarily increased sales and revenue to improve the profitable picture (see strategy II, Figure 3.6). When the costs in strategy I are decreased by 20%, the profit is the same as in strategy II when the sales are increased by 40%. This result is due to the fact that as the revenues increase the associated costs increase in the same proportion. With waste in the system, the waste and cost are carried along with the sales growth. Creating a leaner organization with less waste increases profitability.

Digging deeper into strategy I (Figure 3.7), reducing cost improves profitability and releases cash. The released cash can be banked or distributed, but better yet, reinvested as a fund for innovation for modifying existing products and services, or even creating new products and services. As the products and services improve, there's a high correlation with increased demand and the revenue and market share naturally grows. New products or services increase competitiveness and also influence demand and increased market share. Over time, this strategy optimizes and maximizes business valuation. Value increases, customer base increases, costs decrease, and performance and profitability improves.

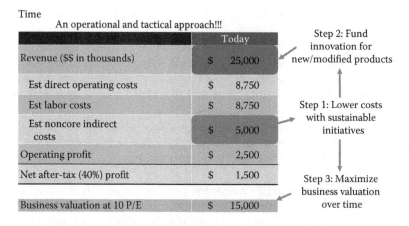

Figure 3.7 Value/cost strategy 1: (1) lower costs, (2) release cash for growth, (3) maximize business valuation over time.

It seems like a *no brainer,* but it's amazing how many organizations are stuck in the old paradigm.

This whole model is built around value and cost. This brings us to the following questions when it comes to value and cost:

- How do you know what is of value?
- How do you know what is the right cost?

Value can be found in what you do, your purpose, your importance, your worth. The right cost resides in the necessary expenses required for you to deliver your products and/or services to your customer. Textbooks tell us that your value add is what you do that the customer is willing to pay for, and everything else is considered nonvalue add. Affordability supports the value-add and nonvalue-add delineation, but also includes and categorizes value-add support as those support functions necessary for delivering value add. In fact, within one project I was involved in several years ago with Dr. Beth Cudney, during the evaluation of value-added functions, value-added support functions, and nonvalue-added functions, we were able to observe, in many cases, about 10% of the organization's entire statement of work spent (people, material, tools, supplies, etc.) was in the value-added area of the system, around 30% of the organization's entire statement of work spent was in the internal value-added support area, and the remaining 60% was in the external value-added support area. Hence, we coined the 10-30-60 rule. Breaking down the value-added and value-added support proportions, we could conclude that often, and typically, only around 10% of the entire organization is applied to the true value add. (Note: The greatest value add we were aware of was Toyota,

Georgetown, KY, where the value add is estimated as high as 30%. So, there is some variation to the 10-30-60 rule.)

Moving from value into cost, we need to investigate the definition of "right cost," the cost of affordability (Figure 3.3) is comprised of the value-added costs and value-added support costs. The "wrong costs" are found in the cost of nonaffordability (Figure 3.3) or those nonvalue-added costs. It is key critical for every organization to have a discussion around value, value-added, value-added support, nonvalue, and nonvalue added. This helps in identifying value-added costs, value-added support costs, and nonvalue-added costs for prioritizing and targeting value improvement and cost reduction.

Often, a challenge exists where the perspective of value and cost gets confused between the customer focus or external perspective, and a value add or internal focus. For example, there is a common type of service where this mixed perspective frequently occurs: healthcare. Traditionally, health care has focused on providing services as centered on physicians, nurses, medical procedures, and medication. Patients are often scheduled around availability of physicians, nurses, and facility availability. This has been the healthcare paradigm that's created so much waiting, and mostly the patient waiting on the doctors, the nurses, the procedures, and the treatment. It's no wonder we've seen elegant waiting rooms emerge over the past number of years. These waiting rooms do serve a purpose for the people waiting for a patient to receive treatment, but they are nonvalue add when it comes to patients waiting long times for the treatment. As has been documented and communicated by the Institute for Healthcare (IHI), the value of health care is healthy communities, the customer is the patient and successful patient outcomes, and the expense and price should be addressed by lowering costs. This is a patient-centered approach. Other than money, time is also a cost of healthcare, and time to treatment can be a life or death situation. From my observations, the three most critical wastes in healthcare are defects, waiting, and motion. Costs can be dramatically reduced by eliminating defects, decreasing waiting, and reducing motion. Using a patient-centered approach, this can be done and the result will add value and decrease cost.

CASE EXAMPLE 1   Healthcare: gastroenterology, procedure—colonoscopy

I'd like to use a simplified, yet real case example of how to use the concepts of value and cost, combined with customer focus, in instituting a value/cost initiative. In Figure 3.8, the customer is a patient, the provider is a gastroenterologist and his staff, the procedure is a colonoscopy, and the process involves a patient visit to receive the procedure, and this qualifies as but one dimension of the patient's healthcare system and one

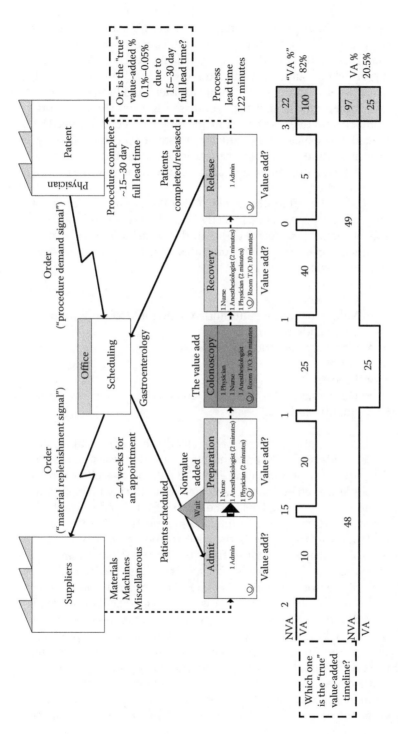

Figure 3.8 Healthcare value and cost example.

aspect of the healthcare enterprise. It's worthy to note that, of the top five deadly cancers, colorectal cancer represents more than 50% of the worldwide cancer deaths each year, depending on whose statistics you're looking at, ranks number 3 or 4, usually behind lung and prostate for men, and lung and breast for women (and sometimes you'll see the top three as lung, stomach, and liver, with the fourth being colorectal followed by breast). Regardless of the ranking, it's safe to say that colorectal cancer qualifies as a major cancer killer. One that can, for the most part, be prevented and successfully treated if discovered and remedied in time.

Using the value stream map (VSM) in Figure 3.8, for value and cost purposes, let us investigate and discover some affordability factors of (keeping in mind, the affordability aim of healthcare is healthy communities, patient outcomes, and reduced cost):

- Customer
- Value (including value added, value-added support, non-value added)
- Cost (expense and price)
- People
- Process
- Performance (in terms of customer satisfaction, time, quality, and cost)

In this example, the customer is clearly the patient, but there are secondary customers who may be individuals related to the patient, or providing care for the patient, as well as others who might even be friends who care about the health and well-being of the patient. This is the focus of our VSM example. It could be argued, from an internal perspective, that the value illustrated in this VSM is the visit to the doctor's office and the colonoscopy procedure from the time the patient arrives until the time the patient is released. It can also be debated that the only reason the patient is there is to receive a colonoscopy and that is the true value of this VSM. (For discussion sake, I included both VA/NVA timelines.) It could also be disputed that the whole value of this VSM is the sequence of preparation–procedure–recovery. I have had a variety of organizations use one of all three. The important point in affordability, value, value added, value-added support, and nonvalue added must be defined, understood, and agreed upon. My personal preference is value added is the procedure, value-added support is the preparation, recovery, and admit/release (admit/release due to government information mandates and standards required by law, which must be included in healthcare procedures and treatment).

When calculating VA% (value add percent), the value add, value-added support, and nonvalue-added dynamics should be clearly defined in order to appropriately focus cost, expense, and price reduction efforts. The VSM in Figure 3.8 depicts three VA%. Each depends on how the organization defines value and value add. (Remember: Value add is what the customer is willing to pay for and nonvalue add is everything else. The target for cost, expense, and price reduction exists in the nonvalue-add areas.) One VA% is 82% and includes all aspects of the procedure, from admittance to release. Another VA% is 20.5% and only uses the procedure as the true value-added portion of the process. VA% can also be calculated from the time the procedure is ordered until the time the patient is released (using this calculation, a lot of discussion can be dedicated to "why does it take so long to schedule the procedure?"). Lastly, and not calculated in Figure 3.8, VSM, preparation–colonoscopy–recovery as VA would yield either VA% of 70% at the procedural level or a VA% of 0.4%–0.2% if calculated from the time the patient needs it until the moment the patient is released. This brings us to the affordability "rule of thumb": increase the value and value-added functions, optimize the value-added support functions, decrease the nonvalue-added activities.

As for cost, money (in terms of price and expense) is not the only cost to consider. Patient satisfaction, provider (doctors, nurses, support staff, even supplier) satisfaction, time, and quality should also be a cost consideration. The efficiency and effectiveness of every process depends on money, satisfaction, time, and quality. Lost customers (patients in this case), dissatisfied people, lost/long times, product/service defects, and missing requirements can all be calculated at some cost, and most even in terms of money.

This is a good time to interject on a statement often used; "Money is the root of all kinds of evil." With this in mind, there's a warning hidden within this discussion of money that I've seen emerge within Lean, Six Sigma, Re-engineering, and even Total Quality Management, all being predecessors, complements, and components of affordability. I have observed many times, with individuals that have hidden and nefarious agendas, the use of all of the methods I mentioned for the sake of cost reduction, but in fact the real agenda was that of people elimination and downsizing. In fact, I witnessed it as a part of one project I was involved in with one of the largest healthcare companies in the United States. I was recruited for the project because of my Lean expertise. But it really turned out to be a cover for identification and eradication of what was termed "useless personnel." To say the least, I found a way to be very busy, and I finished the initial work I committed to and moved on to project with other organizations that sought the true purpose and intent of increasing value and worth through lower cost using problem solving, teamwork, and the people involved in the process without threats of firing and eliminating people.

And finally, speaking of people, the cornerstones, base, and foundation of affordability (see the house of affordability) are comprised of people, process, and performance. All of these factors should be well understood, defined, clear, and apparent throughout all facets of the value stream and system (these are all covered in more detail in Chapters 9 through 11). The people provide the solutions and energy to improve the process. The outcome of the improved process is demonstrated through increased performance. In this case, customer, provider, and supplier satisfaction goes up, time goes down, the process gets faster, quality increases, and the cost decreases.

CASE EXAMPLE 2   Warehousing: subassembly, process—kitting

For my second example regarding value and cost, I'd like to go to an industry and service very different than healthcare. Several years ago, while engaged with several projects of a large electronic components distribution service company, Anixter, Inc., I was involved in one particular project that focused on the process of kitting material for assembly and installation of Cisco System Products (i.e., Internet bridges, routers, switches, etc.) involving two of their partners Jabil and Solectron. This facility was located in St. Petersburg, FL, and served primarily Cisco, Jabil, and Solectron. The Cisco service involved mostly kitting up boxes of material on a per order basis for shipment to Cisco product installation sites around the world. The Solectron service involved the population of kits with material that was a component of the Cisco Products that Solectron assembled. Finally, with Jabil, an electronic board producer, Anixter boxed kits for the pieces and parts used to install the Jabil boards in Cisco products.

The highest paid employee at the facility was the general manager, who was responsible for all services in the warehouse that primarily consisted of fulfilling Cisco, Jabil, and Solectron orders for electronic components and kits. The second highest paid worker was the manager of kitting, who earned around $30.00 per hour and reported to the general manager. The physical layout, not exactly to scale, but close, is illustrated in Figure 3.9. The domain of the kitting operation took up about 10% of the facility's footprint and produced about 35%–40% of the profit since it was a just-in-time, day shift, custom service. There were five such plants across the United States.

Every day, about 8–12 orders would drop in the morning (average of 10) and 8–12 orders would drop in the afternoon (also an average of 10). The way the information system was defined and configured, the kitting manager would receive a notice on his computer that an order was being printed, so he would walk to the printer, pick up the order, and walk back to ready and prepare the kitting team for order fulfillment. The kits would contain anywhere from 4 to 14 items of various materials and instructions. A short time later, the kitting manager had to walk back to the printer to pick up any special instructions or documentation to be included in the kits (note: if he waited for second printing, it delayed the start-up of the order). After the kitting sequence was set up with the material and special instructions and documentation, the kitting process could commence. While the kitting process was taking place, the kitting manager had to go back to the printer and pick up the packing and shipping labels that were printed after the order and instructions. Figure 3.9 somewhat details the motion that took place for each order.

As I observed, documented, and mapped the kitting process, it quickly became clear to me that the process required a lot of motion of the kitting manager. It took the kitting manager about 2 minutes to go to the printer and 2 minutes to return (the 2 minute trip between the kitting manager's office and the printer was the time it took without interruption). For each order, three round-trips to the printer were required. With an average of 4 minutes per round-trip, requiring three round-trips per order, an average of 20 trips each day amounts to 240 minutes walking time. Putting it into perspective, 240 minutes is equivalent to 4 hours, and in a week, the kitting manager, the second highest paid employee, was spending about 20 hours a week walking. For a year's work, that was about 960 hours. I was astounded by the result of the assessment

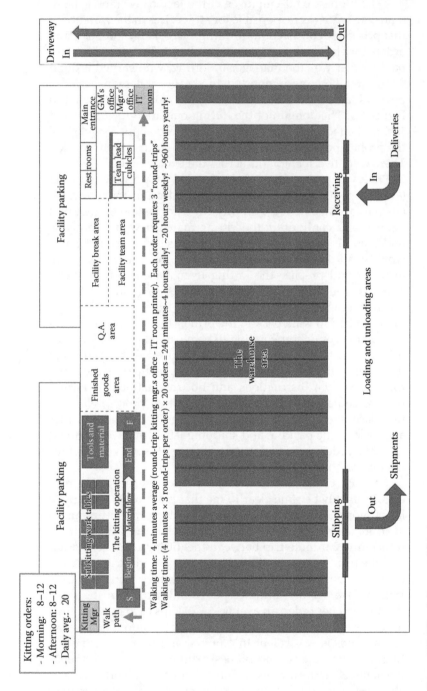

Figure 3.9 Warehouse and kitting operation.

because it meant Anixter was paying approximately $28,800 (960 × $30) to walk. I was almost on the unbelievable side. In fact, with the five locations, configured with the same layout, the grand total came to $144,000.00. Clearly, when considering value and cost, the expense requiring the kitting managers to walk across the plant three times for every order does not fit within the defined value of the kitting manager.

The problem-solving team addressing the issue came up with several solutions. For example, one was to have all of the order information and affiliated documentation drop to the printer and be printed all at once, eliminating two of the three trips (the value/cost calculation still amounted to $9,600.00 yearly for walking). Their final recommendation was to put a printer in the kitting manager's office to provide the printer material directly to the kitting area and kitting team. The recommendation was presented to the vice president of IT, who rejected it on the basis that a cost of $5,000.00 per location would amount to a total $20,000.00 expense for all the facilities combined (the cost included printer, supplies, cabling, and maintenance). Since I had become good friends with the senior vice president of the division, I took the proposal into his office and he quickly approved it.

There is a lesson in this example: even when affordability proves to be the right thing to do; old thinking and archaic business paradigms may still get in the way of superior value/cost proposals.

I do have one related example in the archives that proves that without the proper attention paid to the value and cost dynamics, you could go out of business. It was right around the turn of the century, and I worked briefly with a company called IDS (Internet Distribution Service) that was an upstart Internet company focusing on eBusiness and eShopping. They had a good handle when it came to the value/customer proposition and a good price when it came to customer/cost. However, they failed to balance the value/cost piece of the affordability "puzzle." Although I brought it up to the CEO on numerous occasions, she always had a reason or an excuse that value/customer and customer/cost far outweighed and was more of a priority than expense to deliver the value as it related to the price. The value add of her company was in the performance of the system her customers wanted. Although she paid lip service to that reality, she overspent her funds (provided by angel capital and small investors) on marketing and sales in anticipation of explosive growth. In a nutshell, she overspent in value-added support, underspent in value-added function, created a lot of waste and the deficit grew too great, and her investors shut her down.

In summary, the intersection of value and cost identifies the expense. The expense is in terms of value added, value-added support, and nonvalue added. The key is to increase investment in value added, optimize investment in value-added support, and ever decrease nonvalue-added expense. Always keep in mind, with affordability, expense or cost can be expressed in terms beyond money. Time, quality, and even satisfaction are expressions of expense. Some much too expensive to sacrifice. All of the expense contributes to success, performance, and profitability.

chapter four

Faster

Ease and speed

> The less effort, the faster and more powerful you will be.
>
> **—Bruce Lee**

Faster is about eliminating that which slows you down and increasing that which improves your capability for speed and responsiveness. Faster is about increasing the capability of doing something in less time while increasing the velocity potential of the flow for providing a product or service quicker. Fast and flow go hand in hand. Without impediments, roadblocks, and barriers, flow quickens, speed increases, as well as capacity and capability. We find ourselves in the twenty-first century competing on quality, with lower cost, and the need to deliver products and services with ever-increasing velocity, to meet rapidly growing local, regional, national, and global market demands. Greater speed increases value in terms of responsiveness. Greater speed improves the customer experience, increases loyalty, meeting ever-increasing demands on availability and timeliness. Greater speed also decreases cost as obstacles are removed for swift flow and prompt delivery.

Speed and velocity require that an organization is flexible and focused on meeting their customers' newly emerging and changing demands. While working on a project with the United States Postal Service (USPS), I was fortunate enough to witness a response to a new customer demand. Amazon, now a major customer of the USPS, had a requirement for same day and Sunday delivery. USPS and Amazon then prototyped a service that has proven successful over time. This is how it works: when an Amazon customer orders a product that is in a vicinity logistically capable of delivering same day or next day on Sunday, Amazon can drop the parcel at the local USPS Post Office and that package is delivered as prescribed. With such a new found flexibility, the USPS has now captured more than 40% of Amazon's business and has ultimately proven to outperform both of its major competitors. In fact, with its new level of improvement, the USPS is on the track of affordability. USPS has redis-covered its true purpose and value, improved its customer awareness and responsiveness, while continuing to lower its cost. It is true that there is a long way to go, but the first few steps have been taken, and the future looks bright.

Within affordability, easier and faster go together. In order to go faster, the first step is to make things happen easier. Anything that gets in the way not only slows you down but also creates cost. Easier reduces stress and fatigue of the system. Especially, the stress and fatigue on the people in the process. It's an affordability fact, eliminating people does not make the system easier and faster, nor does it make it better. The first step in making the process easier is to eliminate all those road-blocks and barriers that get in the way. In the last chapter, we discussed a case example of walking in a warehouse. The walking was getting in the way of starting the customer order. It was intuitively obvious that it was easier to kickoff the process by eliminating walking. However, until it was observed, seen, and diagnosed, the problem existed. After removing it, the time from order to delivery decreased and the speed increased.

Have you ever seen the physique of a champion long-distance runner? His body is able to cover long distances at a speed greater than his compe-tition. Although his natural skills are well developed, his gait and stride can be observed as smooth and easy. In Atlanta, during the 4th of July, we have a running event call the "Peachtree Road Race." Thousands of peo-ple participate and as I've watched it several times, it was easy to observe that the runners who labored and struggled did not have a smooth and easy stride. A process is much the same. Processes that operate smooth and easy have greater speed and endurance. The workers have less stress and fatigue. And, when fatigue appears, there is time built-in to rest and recover.

So the question arises: how do you make a process easy and fast? The answer is clear and simple, yet the discovery seemed elusive to many. With the M.I.T. research during the 1980s (for more informa-tion, get the book *The Machine That Change the World* by Womack, Jones, and Roos, Copyright 1990, Published by Harper Perennial 1991), and the implementation of Lean for the past 25 years, we've experienced a revolution of easy and fast. Some organizations still choose to do it the old way, hard and slow. Some of the refusal and reticence resides within the natural human factor of change resistance. Opposition comes from excuses; "L.E.A.N. mean Less Employees Are Needed," "It's a Japanese Method, it won't work here," "What we do is different and unique," "Toyota makes cars, we don't make cars." Our success between

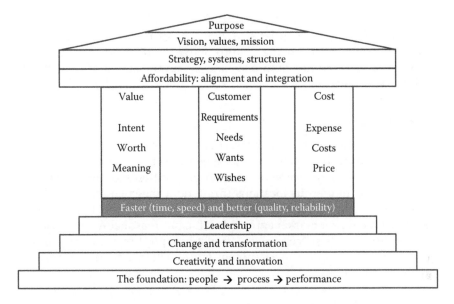

Figure 4.1 The affordability architecture or "the house of affordability."

WWII and the 1980s blinded us to the things that impact easy and fast. Only when the threats appeared, and failure fell upon us, did we take the steps to understand the need for ease and speed. Some still don't get it.

Affordability calls for increased ease and speed to go faster. Lean combined with the quality tools offers a comprehensive toolbox for reducing effort and accelerating velocity. There are numerous performance motives for ease and speed: demand responsiveness, accessibility, shorter lead times, improved flow, efficient fulfillment, and elimination of waste (Figure 4.1).

Demand responsiveness

In this day and age, products and services have to be delivered in accordance with the customers' demand. Today, "I want it when I want it" is often the attitude and demand of consumers. Whether it be cars, books, food, clothing, communication, or service, the instant gratification culture has begun to appear worldwide. In this new global economy, demand responsiveness has emerged as a competitive advantage, perhaps soon becoming the "ante to play" in a product or service industry. Sometimes, this aggressive "pull" for products and services misleads providers into using a bad practice of large inventories. Large inventories of material, or even large inventories of personnel, in a just-in-case philosophy of not being able to deliver and respond to demand. Affordability recommends to pace and synch the speed of the process with the rate of demand.

Accessibility

Following closely with demand responsiveness is accessibility, which also requires speed of delivery and speed of service. Traditional retailers tried to satisfy this type of demand by positioning large amount of inventory and numerous locations throughout the customers' localities. If their gamble and forecast was off, reduced prices and loss offerings resulted. One recent market entry, Amazon, has reasonably responded to this condition quite successfully. Affordability positions and targets specific points accessibility in alignment with customer requirements and expectations.

Shorter lead times

Since the time of Henry Ford, the need and advantage of shorter lead times were known. It is not to say that one should deliver fast and sacrifice quality and cost. It is an "all three" proposal; faster, better, more affordable. By studying process and time, lead times can be reduced.

Improved flow

In keeping with demand responsiveness and accessibility, improved flow increases speed.

The seven flows of manufacturing are material, WIP (subassemblies, partial-assemblies), finished goods, operators/people, machines/tools, information, and engineering. The seven flows as applied to healthcare are patients (including family and relationships), providers (including partners and suppliers), medication, supplies, information, equipment/ machines/instruments, and process/engineering. The seven flows of food service are consumers, food providers (including partners and suppliers), meals, supplies, information, equipment/machines/instruments, and process/engineering. Using the seven flows approach, affordability can be applied to any industry or organization.

Efficient fulfillment

Fulfillment is a philosophy based on consumption and replenishment. Lean refers to this as a "pull system." As something in a process consumes material and/or services, it is replenished by another procedure or activity in the process. Affordability uses this thinking to link together and integrate numerous process modules and procedures.

Elimination of waste

In order to eliminate waste, the barriers and roadblocks that slow the flow must be removed. Anything that gets in the way of process steps and procedures is the target for eradication. Of course, the seven basic wastes (transportation, inventory, motion, waiting, overproduction, overprocessing, defects, memorable as the acronym T.I.M.W.O.O.D.) plus the waste of skill sets (the human factor) are the primary wastes to identify and eliminate to increase ease and accelerate speed. The overall goal is to reduce lead time from when the customer needs it, until the product or service is delivered.

The best way to demonstrate and explain this component of affordability is to present it through the use of two actual case examples. I've chosen two specific examples that I was directly involved with to illustrate and describe how faster and easier can be accomplished even in an environment where speed is not normally an operational condition—government. Both of the examples I have come from a Department of Defense Congressional Program called Mine Resistant Ambush Protected (MRAP). From 2006 to 2013, this program provided combat vehicles to mitigate improvised explosive device (IED) threats in Afghanistan and Iraq. The speed by which this was accomplished was astonishing, the results were even more impressive, and the overall performance was astounding.

Case examples: MRAP vehicles

- Case 1: Integration and deployment
- Case 2: Maintenance and upgrades

MRAP background and purpose

During the beginning of the wars in Iraq and Afghanistan, a deadly threat appeared on the scene known as the IED. From March 2003 through June 2007, the frequency of IED deaths dramatically increased (Figure 4.2).

The majority of the targeted vehicles were troop transportation vehicles known as "HUMVEE" (high mobility multipurpose wheeled vehicle or HMMWV). The root cause of the problem was that the HUMVEE was

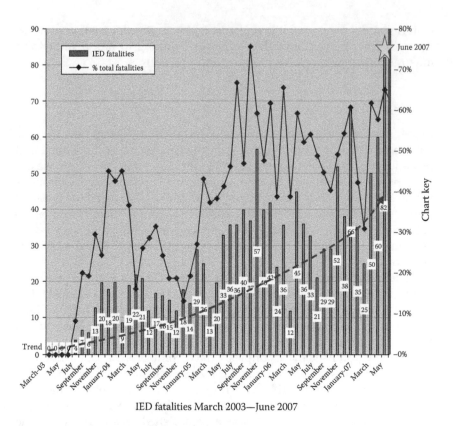

IED fatalities March 2003—June 2007

Figure 4.2 IED threat increases.

The problem	The solution
"Humvee"	Mine resistant ambush protected "MRAP"
❖ High mobility multipurpose wheeled vehicle (HMMWV)	❖ "V" shaped underbody channels blast away from personnel
▪ 52% of all vehicles attacked	
▪ 63% of all casualty producing incidents	❖ Vehicle's height above ground dissipates effects of underbody blasts
▪ 60% of all KIA	
▪ 65% of all WIA	❖ High hardened steel provides armored protection for occupants from fragmentation and small arms fire
❖ Adaptive enemy	
❖ Inexhaustible supply of low tech, highly lethal munitions	

Figure 4.3 IED effects.

not able to withstand IED blasts occurring on the underside of the vehicle. The preferred solution was to provide a fully armored vehicle with a "V"-shaped underbody that could defend against blasts coming from the ground, as well as attacks from insurgents on the ground outside the vehicle (Figure 4.3).

In October of 2006, the MRAP Vehicle Program became the Defense Department's highest-priority acquisition program and the Congress authorized the use of funds to stand up the program ASAP, which delivered the first fully integrated vehicle just months later. The primary purpose of the program was to mitigate the threat of IEDs in Afghanistan and Iraq, and most importantly, Save Lives!

CASE 1: MRAP integration and deployment—SPAWAR, Charleston, SC (March 2007–December 11, 2012)

The design, planning, construction, and kickoff for the integration and deployment facility at the SPAWAR (Space and Naval Warfare Systems Command, Charleston, SC) base began in October 2007 and lasted through February 2007. In March 2007, the facility opened and integrated the first MRAP that was flown from the Charleston Air Base to U.S. Ali Al-Salem Airbase in Kuwait. The magnitude and complexity of the program was quickly realized.

VARIATION

As a target, five of Department of Defense services were targeted to receive MRAPs (i.e., army, marines, navy, air force, special operations command—"SOCOM"). There were six OEM vehicle suppliers manufacturing trucks at six locations across north America (FPII, Summerville SC; BAE, York PA; Armor Holdings, Sealy TX; Navistar, West Point MS; Oshkosh, Oshkosh WI; GDLS, London, Ontario, Canada). Each unique truck model had different designs and dimensions.

The equipment integrated with the vehicles (communications, computers, intelligence, surveillance, and reconnaissance referred to as C4ISR, gun turrets and other equipment) varied from service to service, and the installation process and procedure varied from truck type to truck type (over 100 different truck variants were produced 2007–2012). Eventually, more than 27,000 units were produced. To say the least, it was a very complex condition under which to seek standardization.

On May 2, 2007, the Secretary of Defense, Mr. Robert Gates, distributed a memo, describing the urgency and priority, reiterating the need for speed, and identifying the emerging challenge of variation and complexity. From that day forward, the game was on. Mr. Paul Mann, MRAP Program Manager, was often heard saying, "MRAP is the ultimate team sport!" It turned out, he was right (Figure 4.4).

My first day with the MRAP program occurred on August 20, 2007, at the Charleston, SC, facility. I dedicated myself to spend that week assessing the current condition. That first day, only a few vehicles were completed and shipped to the Charleston Airport. On Wednesday, August 22, 2007, the facility completed the integration of five trucks, and it happened to be the same day the Defense Secretary Mr. Robert Gates promised that 1500 MRAPs will be delivered to the warfighters and/or in route to Kuwait by January 1, 2008 (this was down from the original estimate of 3900 earlier in the year). A commitment

May 2, 2007

TO: DON WINTER, PETE GREEN

CC: GORDON ENGLAND, PETE PACE, KEN KRIEG, TINA JONAS

SUBJECT: MRAP ACQUISITION

Thank you for today's briefing and discussion on the ongoing MRAP acquisition effort.

Allow me to reiterate the fundamental point made during the discussion. The MRAP program should be considered the highest priority Department of Defense acquisition program and any and all options to accelerate the production and fielding of this capability to the theater should be identified, assessed, and applied where feasible. IN THIS REGARD, I WOULD LIKE TO KNOW what funding, material, program, legal or other limits currently constrains the program and the options available to overcome them. THIS SHOULD INCLUDE AN EXAMINATION OF ALL APPLICABLE STATUTORY AUTHORITIES AVAILABLE TO THE SECRETARY OF DEFENSE OR THE PRESIDENT.

I am also concerned with the wide variance in approach on the use of this capability between the Marine Corps and the Army. In this regard, I ask both the Army and the Marines to work with the Joint Staff to reexamine this issue and come back to me quickly on how to field and utilize the added crew protection capability afforded by the MRAP family of vehicles.

It is clear that a lot of good work has been done in getting this program to its current state. However, the urgency of the situation on the ground in the CENTCOM AOR requires that we thoroughly evaluate ALL options to put as much of this enhanced capability in the hands of our troops as rapidly as reasonably possible.

I ask that you get back to me on the specific requests above no later than May 11, 2007.

Signed,

Robert M. Gates (Secretary of Defense)

Figure 4.4 Memorandum: MRAP is the highest priority for the Department of Defense.

that, at the time, seemed to be impossible. The five trucks per day tempo became the de facto baseline for performance. At that rate, being their best performance to date, it would take 900 weeks (working a 6-day week), or about 17.3 years to deploy 27,000 vehicles (with the final delivery occurring circa 2024). Obviously, not nearly fast enough.

In addition, the average truck cycle time was running around 3–5 days per truck or 48–80 hours (the target was less than 1 day per truck or for two shifts working on a truck, or less than a total integration time of 16 hours). The core design of the integration facility allowed for a two-shift workforce to occupy 50 integration bays. With a 6-day work week, at 50 trucks per day, would pace the system at 300 trucks per week, fully integrated and ready for shipment. At that particular rate, 27,000 trucks would take 90 weeks, or about 2 years. Up until this program was created, in the entire history of warfare, no ground combat system has ever gone from concept to fielding of that many units in 3 years ... a clearly impossible task!

Since the need was urgent, the trucks were expedited via airlift from Charleston to Kuwait. This mode of transportation allowed an average of 2–3 MRAPs per plane, with 2 or 3 flights scheduled each day. Since the rate of integration was operating at five or less units per day, this methodology of delivery was more than adequate, but not fast enough to meet the 50 vehicles per day rhythm required.

A strategic view of the situation (see Figure 4.5) illustrates the flow.

The seven traditional flows are present: material, WIP (work in process consisting of subassemblies and unfinished assemblies), finished goods, operators/people, machines/tools, information, and engineering (processes, procedures, and methods). Truck and integration material flow: The first three flows—material, WIP, and finished goods—are the primary value and purpose of the organization. The operators/people perform the integration through the use of machines/tools: installation of communications, computers, intelligence, surveillance and reconnaissance (C4ISR), gun turrets, and other equipment. Information flow throughout, as well as engineering factors of processes, procedures, and methods, is present from the time a truck arrives until it is delivered to the airport or the seaport for transport. Smooth and rapid flow is the ultimate goal.

So, one might ask, how fast is fast enough?

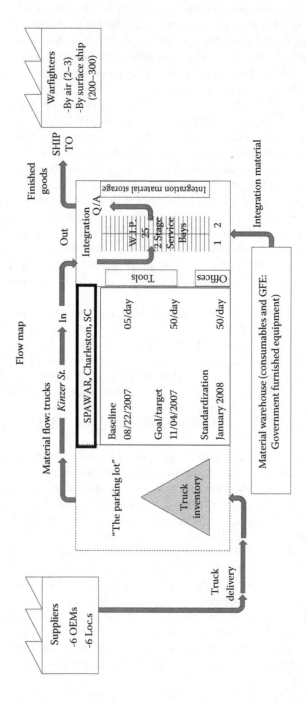

Figure 4.5 Case1: MRAP–integration and deployment.

Since the trucks were being delivered via flatbed trailer from the six OEM production facilities across North America (i.e., the United States had five and Canada had one), it was decided that a parking lot would be set up to hold at least 200 vehicles. This inventory level was deemed acceptable due to the inconsistency and sporadic delivery rate of trucks from the manufacturers. By the end of the first week, it became noticeable that the mechanics performing the integration tasks would walk by the parking lot on the way to work, notice the inventory level, and discuss the amount of work available for that day on the way into the integration facility. Days when there were only a few trucks in the lot, and only a few trucks were being delivered, the work would pace at a rate that matched the number of trucks flowing from the lot and into the lot. This occurred as a result of shifts being sent home early when the availability of trucks to work on ran out. I called this "the parking lot paradigm." A demand was went upstream to the OEMs to "fill the lot with trucks"! This remedied the issue of slow or no flow of trucks for work.

By the beginning of September, with the sense of urgency very high, several Lean efforts were invoked and they all operated in parallel. Since they were not well coordinated, they all seemed to trip over each other trying to do "good things." Finally, by late September, efforts started to amalgamate and a synergy began to emerge.

The main efforts of the Lean implementation initiative were focused on the system's flow and the truck work cells where the integration work took place. As it is with many assembly and integration operation sites, Lean endeavors commonly use 5S, standardization, flow, and pull concepts focused on eliminating waste, and removing obstacles and impediments impacting the performance of the process. This location was no different than most. From September 2007 through December 2007, the Lean effort was primarily focused on (1) 5S the work cells, (2) standardize the work cells, (3) improve the material supply flow, (4) improve the material supply flow, (5) institute a production cell configuration process, and (6) establish a pull system.

5S THE WORK CELLS

From day 1, it was clear that workplace organization was not a critical factor of the operation. When observing the 50 integration cells, it was clear that no two were alike, and the practices used for cell organization and cleanliness widely varied.

Some of the work areas were referred to by others as "pig pens." Most cells were incomplete when it came to tools and toolsets, and often borrowed tools were never returned, causing mechanics to spend a lot of time searching for tools and toolsets. There was a shortage of machines, as well as a lack of standard procedures and methods. September and a large part of October was spent organizing the work cells and discovering best practices.

STANDARDIZE THE WORK CELLS

Once the work teams were able to identify and institute best practices, a lot of attention was paid to point of use (POU) for each cell. Cell content, location, and positioning of tools, materials, and machines, with standardized practices and procedures, were emphasized by the continuous 5S effort. Eventually, standard cell designs emerged and were amply outfitted with everything necessary to perform the work requirements of integration.

IMPROVE THE MATERIAL SUPPLY FLOW

Beyond the slow flow of trucks, material shortages and a smooth flow of supplies had to be addressed. What was once delivered as "pieces and parts" became delivered as prekitted configuration ready material carts containing everything that was needed for a particular vehicle variant and configuration. As a truck moved into a cell, all the material needed for that truck was rolled up on a prekitted cart.

INSTITUTE A PRODUCTION CELL CONFIGURATION PROCESS

Each cell was prestaged for next vehicle that was scheduled to arrive. This eliminated the time of waiting for vehicles, materials, and supplies. As a truck arrived, the mechanics knew what to do and everything was delivered that was required for that vehicle.

ESTABLISH A PULL SYSTEM

As a truck moved out of the integration facility as finished goods, ready to be flown or shipped to Kuwait, another moved into place to be verified and validated for shipment in the Q/A area. Movement of trucks as finished goods out of the Q/A area permitted other trucks from cell position 2 to move into place for validation/verification. Trucks in position 1 moved into position 2, and trucks from the parking lot are moved into their appropriate position 1 designation. As trucks flow out of the parking lot, the OEMs can position their trucks in their section, ready for integration.

It was a fine thing to witness the first day the site completed 50 trucks on November 4, 2007. January 2008 was the first entire month that the rate of 50 a day was accomplished. Later in 2008, the record rate of 75 in one day was achieved. How fast was "fast enough"? 50 a day (Figure 4.6).

VALUE STREAM MAP

From the beginning of the program, a value stream map (VSM) was used to illustrate the enterprise flow of the program. This visual permitted program members, numbering in the hundreds, to see the strategy, measurement, information, and movement of material from the point of acquisition to integration, to delivery. Over the years, the VSM practice was updated and maintained until integration was complete. For logistics purposes, the U.S. Transportation Command (TRANSCOM) directing the movement of vehicles using the USAF airlift resources that paced traffic at 2–3 MRAPs per plane in 2–3 days, and Military Sealift Command (MSCO) that operated at a rate of 200–300 trucks (average 224) per ship from Charleston to Kuwait in 21–23 days. From November 2007 through the end of the year, the speed and quickness of integrated vehicle delivery accelerated dramatically. By mid-December 2007, both air and sea logistics were in place and in full action. The article later, featured in "Stars and Stripes" magazine, best tells the story in the midst of the surge of momentum (Figure 4.7).

There was a magnificent team in action and an outstanding effort in place. It was a great pleasure working with such an excellent organization. There's nothing like facts-based personal testimony for any proof of concept. Two articles published in January 18, 2008 (http://archive.defense.gov.news), describe the status and emerging results during the first phase of deployment. They also serve to describe affordability at its best. Imagine, a U.S. government organization responding with urgency, haste, hustle, and speed!

ARTICLE 1: "GATES OBSERVES MRAP PROGRESS, PRAISES WORKERS BEHIND IT"

(Source: U.S. Department of Defense; issued January 18, 2008)
http://archive.defense.gov/news/newsarticle.aspx?id=48709

Charleston, SC: Defense Secretary Robert M. Gates visited here to see progress made in speeding up delivery of

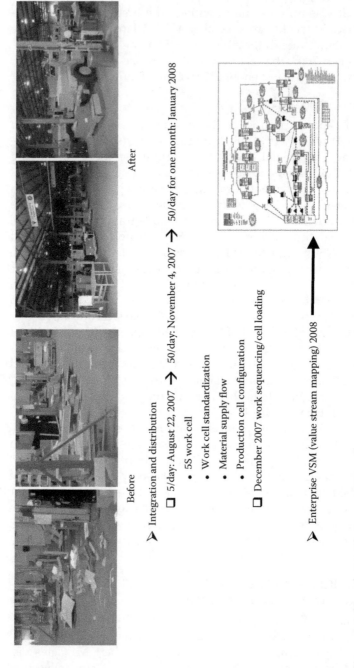

Before

After

➢ Integration and distribution

☐ 5/day: August 22, 2007 → 50/day: November 4, 2007 → 50/day for one month: January 2008

- 5S work cell
- Work cell standardization
- Material supply flow
- Production cell configuration

☐ December 2007 work sequencing/cell loading

➢ Enterprise VSM (value stream mapping) 2008

Figure 4.6 How fast is "fast enough?"

DOD is expected to surpass MRAP goal on Thursday

By Jeff Schogol
Stars and Stripes
Published: December 18, 2007

ARLINGTON, Va.—More than 1500 Mine Resistant Ambush Resistant vehicles are expected to be downrange as of Thursday, Pentagon spokesman Geoff Morrell said.

The MRAPs have V-shaped hulls that deflect blasts from underneath, providing better protection than up-armored Humvees. But MRAP vehicles also are often too large to go off-road, into confined spaces and cross bridges.

Defense officials had said the department's goal was to deliver 1500 MRAP vehicles to the U.S. Central Command theater of operations by the end of the year.

"Our hope is on or about Dec. 20, we will exceed that number," Morrell told reporters Wednesday.

Defense officials have delivered 1330 MRAPs to theater as of December 17, with another 180 vehicles enroute by sea and 15 more vehicles being airlifted downrange, he said.

Morrell said not all of the vehicles delivered downrange will be fielded to troops by the end of the year, it would be close to 1500.

Morrell's comments came one day after Defense officials placed their latest order for MRAP vehicles, asking for 3126 more of the heavily armored vehicles. Those would be in addition to the 8800 vehicles already under contract this year.

In October, a senior Defense official told reporters that the Defense Department planned to order 6500 of the vehicles in December.

Asked if the Defense Department had decided to curtail MRAP orders, Morrell said no and noted that commanders in Afghanistan have requested more.

The plan is still to get more than 15,000 MRAP vehicles to all branches of the service, and that number could increase, Morrell said.

However, the fervor surrounding MRAPs seems to have cooled since earlier this year.

In March, Marine Corps Commandant Gen. James Conway told reporters it was a moral imperative to get all Marines and sailors going outside the wire in Iraq into MRAPs. In May, Defense Secretary Robert Gates wrote a memo calling the MRAP program "the highest priority of the Department of Defense acquisition program."

But at the end of November, the Corps scaled back its request for MRAPs due to the vehicles' limitations and the improved security situation in Iraq.

Last week, the head of Multi-National Corps – Iraq said the Army would need fewer MRAPs after U.S. troop strength in Iraq falls next year.

Still, Morrell said Wednesday that MRAPs are still needed downrange.

"Despite whatever the limitations there might be on these vehicles, they are proving to be extraordinarily valuable life-saving, and the commanders in Afghanistan seem to want more of them."

Figure 4.7 MRAP: speed to service.

mine-resistant, ambush-protected vehicles to warfighters and to thank the people working behind the scenes to save military lives.

The secretary toured the Space and Naval Warfare Systems Center, a massive warehouse where teams crawled in, on, and around nearly 60 of the massive MRAP vehicles on the factory floor, installing radios, sensors, jammers, and other equipment.

The crews, who work around the clock in two shifts equipping more than 50 MRAPs a day, crowded around a podium and took places on MRAPs flanking it as Gates compared them to their World War II–era predecessors.

He cited President Franklin D. Roosevelt, who called on the production lines to "raise their sights" and proved wrong anyone who said that what they were striving to achieve couldn't be done.

"Those in the MRAP program have shown that it can be done. So keep raising your sights. Keep these vehicles rolling off the line," Gates said. "Your efforts are saving lives."

Gates called the MRAP "a proven lifesaver on the battlefield" that helps protect against cheap, deadly and difficult-to-detect improvised explosive devices, which have been the no. 1 troop killer in Iraq. He noted that in 12 of the Army's deployed MRAPs that have come under attack every soldier aboard walked away.

The secretary shared a deployed sergeant major's description of the MRAP as "just lovely," drawing laughter when he admitted that the soldier's actual words were "considerably more colorful."

Gates went on to share more of the sergeant major's assessment: "Troops love them. Commanders sleep better knowing the troops have them."

"There can be no better description of the difference you are making here. You are saving lives," the secretary told the workers as he extended his thanks, along with those of "countless moms and dads, husbands and wives, and sons and daughters of U.S. troops deployed abroad."

Gates conceded that there is no fail-safe measure to prevent all loss of life and limb on the battlefield. "That is the brutal reality of war," he said. "But vehicles like MRAP, combined with the right tactics, techniques, and procedures, provide the best protection available against these attacks."

Since the secretary made MRAPs the Defense Department's top acquisition priority, the program has advanced at near-unprecedented speed. The department met its year-end goal of getting 1500 MRAPs to the theater, and by January 16 had delivered 2225 MRAPs.

"The last time American industry moved from concept to full-rate military production in less than a year was World War II," Gates said.

He cited a monumental partnership between government and industry and the willingness of workers like those at the Space and Naval Warfare Systems Center to work around the clock 6 days a week to meet requirements.

"I don't think it will surprise you to hear me say you must keep pressing on," Gates told the workers. "IEDs will be with us for some time to come in Iraq, Afghanistan, and battlefields of the future. The need for these vehicles will not soon go away."

Gates walked through a static display of nine MRAP variants for all four services and rode past a huge staging area where more than 300 vehicles were about to be loaded for shipment to the theater.

From there, Gates went to Charleston Air Force Base to watch an MRAP destined for Army Special Operations Command in Iraq being loaded aboard a C-17 Globemaster III cargo aircraft.

Air Force Maj. Chad Morris, 437th Aerial Port Squadron commander, described operations that have airlifted 1609 MRAPs to the theater to date. "Our target was 360 a month, but we're pushing that out," he said.

"Our folks are here, committed to the mission and doing the job," Morris said he told Gates. "We haven't missed a beat yet."

On the factory floor and tarmac, workers and airmen alike said they were happy to show the secretary how they are contributing to the MRAP mission.

"It's not every day we get a man of his integrity to come out and see us," said Andrew Harkleroad, who has worked at the center for 15 months. As an Air Force Reserve staff sergeant, Harkleroad said he has a personal interest in knowing that the country is pulling out all stops to protect its deployed service members.

"It's an honor for him to come down here and talk to us," agreed Andrew Fuentes, another center employee. Fuentes said he gets a lot of gratification knowing he's supporting the troops. "I feel I'm participating. I'm saving lives," he said. "That's the objective."

At the air force base, Air Force Tech. Sgt. Eugene Porter called Gates' visit "awesome."

"It shows our men and women in the military we're doing what we can to support our comrades overseas," said Porter, a shift supervisor for the air terminal operations center. "We're getting our stuff to the fight, and we're doing it all as quickly as we possibly can."

ARTICLE 2: "CHARLESTON OPERATIONS SPEED MRAPS TO THEATER"

(*Source: U.S. Department of Defense; issued January 18, 2008*)
http://archive.defense.gov/news/newsarticle.aspx?id=48716

Charleston, SC: It's a typical day in this charming southern city: cobblestone downtown streets swarm with tourists, magnolias are about to pop, and hundreds of mine-resistant, ambush-protected vehicles are being readied for transport to troops in Iraq and Afghanistan.

The glossy travel brochures might not note it, but Charleston has become the epicenter of a massive Defense Department program to get more heavily protective vehicles to deployed troops.

Defense Secretary Robert M. Gates got a firsthand look today at operations under way to speed up delivery of MRAPs to the combat theaters through a program he moved to the fast track in summer 2007.

The vehicles, with V-shaped hulls that help deflect underbelly blasts, have proven to be lifesavers against improvised explosive devices and the even-more-deadly explosively formed penetrators.

After visiting Space and Naval Warfare Systems Center Charleston, where radios, sensors, and jammers are installed in the vehicles, Gates watched as an MRAP bound for the war zone rolled onto a C-17 Globemaster III aircraft at Charleston Air Force Base.

The Defense Department initially flew all MRAPs to the theater as soon as they were ready, and the 437th Aerial Port Squadron here continues airlifting an average of 12 vehicles a day, said Air Force Capt. Jim Lovell, the squadron's flight commander. Since getting the MRAP mission in April 2007, the 437th has delivered 1571 to Iraq and Afghanistan.

They're flown aboard Air Force C-17 Globemaster III and C-5 Galaxy aircraft that can carry about three MRAPs at a time or on contracted transport planes capable of flying up to six, Lovell said.

Several miles away, along the banks of the Cooper River, another MRAP delivery mission is under way and quickly makes gains on the airlift effort. The Army's 841st Transportation Battalion, based at Charleston Naval Weapons Station, is overseeing a sealift operation capable of moving 200 to 300 MRAPs at a clip.

Shipped aboard a Military Sealift Command ship—specifically, large medium-speed, roll-on/roll-off vessels, known as LMSRs—or commercial cargo ships, the MRAPs take 21–23 days transit time to reach Kuwait. They're offloaded there and moved north into Iraq.

The sealift effort began in the fall, when the 841st shipped 48 MRAPs in early November, 180 later that month, and almost 550 in December. "That's certainly a success story," said Army Maj. Isabel Geiger, the 841st Transportation Command's operations officer.

Army Lt. Col. Randolph Haufe, the unit commander, said the pace of shipments will only go up as the production line produces more vehicles.

Army Master Sgt. Kevin Young, the unit's operations non-commissioned officer, described the loading operation that moves hundreds of vehicles onto the ship and secures them for transit within the span of 6–8 hours as "a pretty dance, with everyone working in tune."

Because MRAPs come from several manufacturers and in a variety of configurations, there is no one-size-fits-all way to load them, explained Craig Messervy, a marine cargo specialist and Army veteran. Messervy spends hours formulating a load plan that gets as many MRAPs as possible onto a ship while ensuring a safe passage and no damage to the vehicles.

As he watches, each vehicle roll up the loading ramp and into position, Young said he likes to "give it a little pat" of encouragement and gratitude for the service it will provide his fellow soldiers.

Haufe said it's no coincidence that Charleston has become the hub for MRAP shipments. The largest MRAP producer, Force Protection Inc., is just a few miles down the road in Ladson, S.C. The Space and Naval Warfare Systems Command here already had been integrating electronic systems into uparmored Humvees when the MRAP mission kicked into high gear this past summer.

But even more significantly, Haufe said, Charleston is "DoD's premier intermodal hub" with almost unparalleled transportation assets. It boasts a railhead, an Air Force Base with a C-17 wing, and a secure port facility that's able to spill its operations to the North Charleston Terminal directly next door when the need arises.

"This is one of the few secure DoD facilities that has the capability to move general cargo in and out in large quantities," Haufe said.

And because the Navy owns the terminal, the government saves berthing fees that can run as high as a quarter million dollars for an LMSR ship.

"You combine everything—the added security, the fact that we're a military town, the intermodal hub, and a great, experienced workforce—and this is just a terrific place for the MRAP," Haufe said. "It's no accident that we're shipping the MRAP out of Charleston."

Everyone involved in the MRAP delivery mission, whether by air or sea, agrees they're happy to carry out a mission that supports deployed troops so directly.

For Lovell, the mission is deeply personal. He lost a family member, a Navy corpsman, to an enemy IED in Iraq's Anbar province. Lovell said he's convinced that if his relative had been in an MRAP rather than an uparmored Humvee, he might have survived.

"We all have a vested interest in getting these out on time, and we're doing everything we can to protect these men and women in uniform," Lovell said. "What we're doing is not for show. It's making a difference."

Army Sgt. 1st Class Frederick Jones, operations non-commissioned officer in charge for the 841st Transportation Battalion, said he witnessed the MRAP's capabilities firsthand while deployed to Tikrit, Iraq. The first "Buffalo" model MRAPs had just arrived in the theater, and troops initially didn't know what to make of the vehicles that have been referred to as "Humvees on steroids."

But after seeing both standard uparmored Humvees and MRAPs come back from missions after being hit by IEDs, the troops were sold. "I saw the difference MRAPs make, and it's a big difference," Jones said.

Air Force Staff Sgt. Frank Douglass, load team chief for the 437th Aerial Port Squadron, had just returned from a deployment to Baghdad in April when he was assigned to the MRAP airlift mission. "I came home and jumped right into it," he said. "Soldiers could die if these don't get there. We know that, so we do everything we can to load them correctly and make sure nothing goes wrong in flight so they can get downrange to the people who need them."

Air Force Capt. Ruth Meloeny, a C-17 pilot with Charleston Air Force Base's 16th Airlift Squadron, said flying MRAPs is little different from flying toilet paper, bubble wrap or any other commodity—at least until the cargo ramp drops at the destination. "You really see a difference," she said. "When guys see those MRAPs roll off, they're very happy. You've brought them something that's giving them increased safety and helping making them more effective in doing their job. That's pretty rewarding."

Many of the civilian stevedores and longshoreman who represent the heart of the sealift operation are veterans themselves and see the mission as a way to continue serving. Among them is retired Army Sgt. 1st Class Kim Green, former NCOIC of the 841st Transportation Battalion and now a civilian marine cargo specialist supporting his old unit.

"Our fellow soldiers over there are at it 24-7, so this is the least we can do to show we're behind them," Green said. "We're still contributing. We're still serving."

Haufe attributes much of the success of the sealift mission to a workforce that's willing to put in the hours required—including most weekends—to keep up with the requirement.

"The vessels just don't stop, but they understand the importance of the MRAP and how much it contributes to the safety of our soldiers, sailors, airmen and Marines," he said. "We're all proud to be moving the MRAP. That thing saves lives, and we're happy to be playing a part."

The result of the rapid deployment of MRAPs was extraordinary. From the program's inception in October 2006 through the halfway point in fielding, April 2009, the snapshots in Figure 4.8 illustrate the conditions as deployment ramped up, until it reached its midway point. After April 2009, until the end of deployment, the rate integration and the speed of distribution and logistics seemed almost easy and effortless. The availability of the MRAP in the war theater saved thousands of lives, and the outcome was heralded as the fastest fielding of a ground combat system, from concept to completion, of all time.

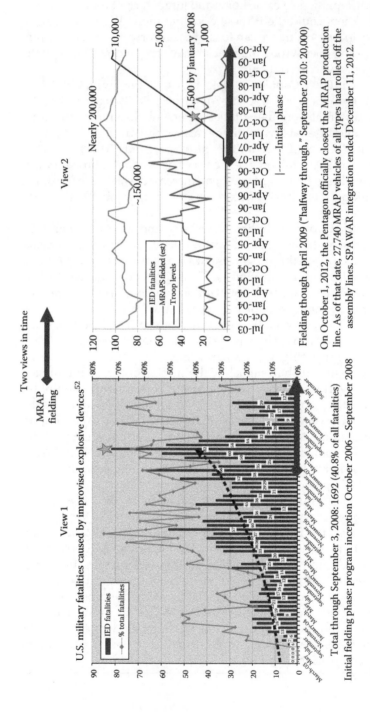

Figure 4.8 MRAP overall results: IED fatalities reduced.

CASE 2: MRAP maintenance and upgrades— Afghanistan (October 2009–April 2011)

After the initial phase of deployment (FY2007–FY2008, October 2006–September 2008), and during the completion of the primary distribution of more than 20,000 vehicles (FY2009–FY2010, October 2009–September 2010), and as a result of their frequent use and utilization in combat, the demand for the speed of faster battle damage and repair (aka: BDAR) increased, as well as a need for increasing the rate and velocity of upgrades for the vehicles requiring improvements. Maintenance and upgrades of the vehicles in the fight were all done in the war theater (Afghanistan, Iraq, Kuwait) at various bases throughout the three countries. During FY2009 (October 208–September 2009) demand for BDAR and upgrades had increased to a level so great that it far outstripped capacity and capability and jeopardized the safety of the warfighters because in order to keep on fighting during repair and upgrade time they had to substitute their MRAPs with uparmored HUMVEEs. A program snapshot was taken at the beginning of October 2009 (see Figure 4.9) that illustrated, although deployment was coming to an end, maintenance and upgrades were becoming the primary focus of the

---

MRAP FY2010 (Govt. Fiscal Year 2010: October 2009 – September 2010) Highlights and Needs

- DoD's Largest $ Program in FY10
  - Larger than the Missile Defense Agency
- ~ 26,600 Vehicles Procured in Less Than 3 Years (FY2007, FY2008, FY2009)
  - Concurrent Production, Testing, and Fielding
- ~ 22,600 Vehicles Fielded to Warfighters in Theater
- Operational Readiness Remains > 90%
- ~ $44.1B Total Appropriated Through FY11
  - OCO Supplemental Funding
- 90+ Contracts, 6 OEMs, 100 + Variants
- FY2010 Need: MRAP BDAR (Battle Damage and Repair)
  - Cycle Time and Speed Critical
  - Standard approach and process required
- FY2010 Need: Upgrades
  - M-ATV Underbody Improvement Kit (UIK) Being Fielded in Theater
  - JUONS requirement is 4546, Ramping up to 400+ kits installed/month
  - Ongoing Upgrades for MRAPs other than M-ATV

Figure 4.9 MRAP October 2009 (start of FY2010) snapshot.

program. After 3 years of deployment, MRAP could claim that it became the fastest fielding of any ground combat system in the history of the world from concept to deployment! But with the urgency level of repair and upgrades demands, there was no time to stop and celebrate. The lives of U.S. warfighters were at stake, and the aim of the program shifted from deployment to maintenance. With a new program manager, Mr. Dave Hansen, the aim, focus, and effort of the program swung from production, integration, and deployment to maintenance, upgrades, and sustainment.

It is note worthy that, from the beginning of the program, a conscious decision was made to proactively distribute repair parts in anticipation of an increase in demand that was eventually going to happen. Much of this decision occurred as a result of the state and condition of logistics and technology on the ground in those war-torn countries. The logistics for Afghanistan was archaic and antiquated, much of the logistics of Iraq in the battle areas had been greatly destroyed, and logistics from Kuwait stretched the supply lines. The flow of parts and materials was accelerated and prematurely outpaced the rate of repair (note: this came under criticism initially, but was ultimately heralded as a great strategy using intelligent foresight). With parts being available, facilities being built, personnel being trained, equipment and tools being delivered, information systems being improved and installed, only one thing stood in the way of success: a reengineered process that operated fast enough to meet accelerated demand.

As is in all affordability success stories, an assessment and analysis was executed, a design and plan was put in place, and a team was assembled to address, support, and engage in process improvement on-site (Figure 4.10).

The three primary locations targeted in Afghanistan for process improvement, stabilization, and standardization were (in order of the plan and schedule) Camp Leatherneck in the Helmut Province, Kandahar in the Kabul Province, and Camp Bagram in the Parwan Province. An advance team was sent forward in April to assess the current condition and determine what should be done to improve the process. Lt. Col. Brian Fulks USMC was assigned as the lead officer for the Lean Six Sigma Support Team (called "LSS team") organized to bring the Lean expertise to each site and work with the repair personnel for process and performance improvement.

Figure 4.10 Camp leatherneck, Helmand province, Afghanistan BDAR and upgrade improvement team.

The first and foremost task was to establish good rapport and teamwork throughout the MRAP maintenance location. It was a major challenge since the mechanics or Field Service Representatives (FSRs) of each OEM had separate sections and domains in the service repair buildings called K-SPANs. There were three K-SPANs containing at least one OEM group each. Every set of OEM mechanics used a different approach for BDAR. Every one of the three K-SPANs was organized differently. The culture of every team of mechanics was unique.

From the preliminary assessment, and our own assessment, there were three specific needs of focus (other than unification and teamwork): workplace organization, material flow, waste elimination, and process standardization. The first task was to organize each K-SPAN using the 5S technique. This increased the space utilization. The K-SPANs were repairing 5–6 trucks at any time. After the 5S event in each building, the capacity increased to a 10–12 level.

The next activity was to leverage the momentum and motivation gained by the 5S event and focus on material flow. This permitted the team to "5S" the rest of the location and organize the flow of trucks and materials. Once the flow patterns were established, each repair bay area inside the K-SPANs were better organized using POU philosophy to locate the appropriate equipment, tools, and materials necessary for servicing the vehicles for BDAR and upgrade activities.

The next focal point was on the incoming trucks. Each truck that arrived was coded and parked in a manner of priority and sequence that permitted easy access and retrieval for repair. As trucks were repaired and upgraded, the next truck could be pulled into the appropriate K-SPAN for service. Using a combined organization, flow, P.O.U, and pull approach, the process speed increased dramatically.

Finally, attention was paid to the repair process in terms of stabilization, standardization, and sustainment. The sequence of assess, design, implement, and maintain was used to (1) understand the problem and its root cause, (2) determine a solution, (3) apply the solution, and (4) maintain the process in a consistent and effective manner. The entire month of June 2010 was spent in Camp Leatherneck prototyping and instituting the improvements. The results were notable (Figure 4.11).

Upon completion, the LSS team went to Kandahar, then Bagram and performed the same function using the same approach. Within 1 year, the speed of service had doubled, and the capacity and capability had also more than doubled. Since the resources used on the LSS team combined with the resources already available on each base, the cost of the effort had already been covered and the improvements and gains were accomplished without any additional expense. In 2011, the summary was presented to the program manager Mr. Dave Hansen (Figure 4.12).

With the involvement of thousands of individuals worldwide, and the use of many more thousands of warfighters, the MRAP program can claim victory. It was the "Ultimate Team Sport"!

The MRAP affordability summary

Value: How much would you be willing to pay to protect the lives of six warfighters worth approximately $3,000,000.00 total cost to the Department of Defense? More than money could buy!

Customer: The requirement—IED mitigation to save lives. Proven!

Cost: Purchase price + repair/maintenance equipment + transportation + manpower < $2,000,000.00 per vehicle.

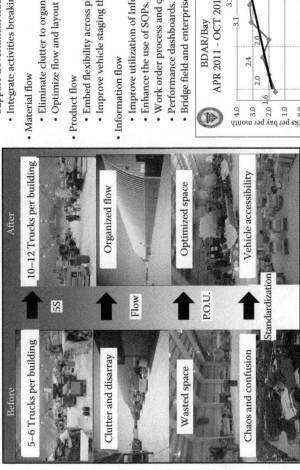

- Teamwork: one-team one-fight
  - Collaborate with on-site personnel as a team.
  - Reinforce chain of command.
  - Support Government leadership.
  - Integrate activities breaking down silos.
- Material flow
  - Eliminate clutter to organize for success.
  - Optimize flow and layout for extra capacity.
- Product flow
  - Embed flexibility across platform and mission.
  - Improve vehicle staging throughput.
- Information flow
  - Improve utilization of Information System and implement best practices.
  - Enhance the use of SOPs.
  - Work order process and quality alert procedure improvement.
  - Performance dashboards.
  - Bridge field and enterprise perspectives.

Results delivered:
✓ Doubled capacity
✓ Improved throughput
✓ Increased efficiency
✓ Increased effectiveness

BDAR/Bay
APR 2011 - OCT 2011

Trucks per bay per month

4.0
3.0
2.0
1.0
0.0

3.2
3.1
2.8
2.4
2.0
2.0
1.6

Baseline   R.I.E.   Result
APR MAY JUN JUL AUG SEP OCT

After
10–12 Trucks per building
Organized flow
Optimized space
Vehicle accessibility

5S
Flow
P.O.U.
Standardization

Before
5–6 Trucks per building
Clutter and disarray
Wasted space
Chaos and confusion

Figure 4.11 MRAP: BDAR and upgrades. Results and outcomes.

Situation	Solution

Leatherneck, Afghanistan Targets
– Improve BDAR Cycle Time
– Increase UIK2 Cycle Time and Capacity
– Optimize Truck/Material Flow Capability
– Improve Information Flow/Utilization
 	
Outcomes:
– Intent: "Double Capacity, Double Performance"
– Reduced/Eliminated Waste
– Faster and Better
	The primary approach and framework utilized was Lean Six Sigma:
	  – Process Cycle Time Improvement
	  – Lean Design Constructs
	  – Material and Information Flow Concepts
	  – Tooling and Technique Implementation
	  – Methods (Point-of-Use, 5S, Standard Work, Cycle Time)

Products and Deliverables	**Results and Impact**

– Facility Capacity Increase
– Site Organization and Capability
– Vehicle Accessibility
– Improved Vehicles per Bay per Month Performance
– Process Performance Improvement
– Customer Satisfaction Results
  – Site 1: Camp Leatherneck
  – Site 2: Kandahar
  – Site 3: Bagram
– Ongoing use and utilization of LSS Resources
	Both the BDAR and UIK2 backlog were addressed and the delay in MRAP availability for the Warfighter was reduced:
	  – Time (Speed to Solution)
	  – Standard Work (Consistent Reliable, Repeatable Methods)
	  – Motivation of the Workforce (Worker Defined Solutions)
	Directly enabled accomplishment of Goals 4 and 5 of the JMVP Strategic Plan.
	All done with existing resources. At No additional cost!

Figure 4.12 MRAP: BDAR and upgrades. Quad chart summary of achievements.

This truck survived an IED attack and protected the lives of six American warfighters.

Figure 4.13 MRAP overall results: lives saved! (Note: For additional information research: MRAP.)

> Expense: Of the $55,000,000,000.00 total allocation for the program, only $52,000,000,000.00 was spent and $3,000,000,000.00 was returned to the government coffer. Expense and price—worth it!
> *Bottom line*: Fast—ease, speed, and velocity. Fastest ever!

One of the most memorable artifacts that tells the story of the program in one snapshot was archived in 2012 (Figure 4.13).

The program was officially ended September 30, 2013, after all MRAP assets were transferred to their respectable services for ownership and sustainment of every remaining unit.

chapter five

Better
Quality and capability

> The bitterness of poor quality remains long after the
> sweetness of low price is forgotten.
> Without continual growth and progress, such
> words as improvement, achievement, and success
> have no meaning.
>
> **—Benjamin Franklin**

Back in the early 1990s, the Bell Labs American Archetype Research was
revealed to me by Lewis Hatala (coauthor of *Incredibly American*). One of
the discoveries of that research focused on quality and quality archetypes
of different country cultures. At that time of the research study, three
countries dominated production throughout the world: Japan, Germany,
and the United States. It was discovered the Japanese archetype for quality
was zero defects of "perfection." The German archetype for quality was
"precision." The American archetype for quality was "perception." While
studying the various quality archetypes, I was able to define a method for
defining quality that fits all archetypes.

I view quality from three perspectives:

1. Conformance to requirements (the customer perspective)
2. Compliance with standards (the value and cost perspectives)
3. Continuous improvement (the strategic growth and sustainment
   perspectives)

Conformance to requirements, compliance with standards, and continuous improvement coincide with each of those major archetypes of perfection, precision, and perception. Too often, organizations frame quality in
terms of their defects, or their lack of defects. This seems to be a negative
way of looking at quality performance. Although measuring and mitigating defects are part of a good quality system at the operational and tactical levels, conformance, compliance, and continuous improvement are at
the positive strategic approach level. Conformance and compliance aligns
affordability's value and customer through requirements and standards.
Continuous improvement also aligns value and customer, with cost in

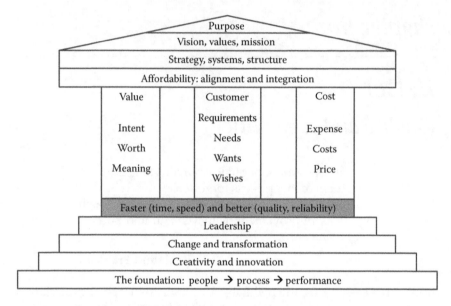

Figure 5.1 The affordability architecture or "the house of affordability."

terms of expense and price. Conformance, compliance, and continuous improvement are the mortars that bind the faster and better elements holding up the three pillars of the house of affordability (Figure 5.1).

Knowing the detail of the customers' requirements, the market requirements, and the industry standards (both established and "de facto"), conformance, compliance, and continuous improvement provide the prescription and direction for excellence in quality, ample capacity, and competitive capability. Chapter 6 provides a detailed approach for developing a strategic plan to meet and exceed customer expectations, surpass market requirements, and deliver a competitive offering that beats the competition. This includes the speed, availability, and responsiveness discussed in Chapter 5. Using a cause-and-effect diagram approach, several factors can be considered for defining and creating a resilient quality system. For designing and developing an excellence quality system, an Ishikawa or "Fishbone" diagram can be created out of eight causes amounting to excellence in quality:

- *Manpower*: The skills and abilities of the people and human resources necessary to accomplish the design
- *Methods*: The methods including the processes and procedures to be employed
- *Materials*: The supplies and resources required that support and institute the system

- *Machines*: The tools, instruments, and machines required
- *Money*: The capital necessary including what funds and where the money should be spent
- *Management*: The type of management and leadership needed to employ the system
- *Mother nature*: The environment expected and assumed
- *Miscellaneous*: Anything else required

When designing a quality system, consideration should be taken to clearly define the strategic quality dimensions.

Conformance to requirements:

- *Customer*: What the customer wants, needs, and wishes
- *Market*: What the market place requires
- *Regulation*: What the industry demands

Compliance to standards:

- *Stabilize*: Stability at the process level, system level, and enterprise level
- *Standardize*: establish a discipline of standardization and standards
- *Sustain*: incorporate cultural behaviors for maintaining process performance, improving process performance, and increasing process performance

Continuous improvement

- *Assess*: Assess the current condition and illustrate the current state
- *Design*: Design and plan for implementing the future state
- *Implement*: Implement the design and plan for improvement(s)
- *Maintain*: Maintain and sustain the process and system performance
- Tools to use include (not limited to)
  - *Designing*: VOC/HOQ/QFD
  - *Operation*: VA/VE
  - *Sustainment*: FMEA

A key factor for process performance improvement is to increase overall effectiveness and efficiency by eliminating conditions that produce defects and cause excessive variation. The most cost-effective and inexpensive approach is to insure quality by designing a defect-free, low-variation product or service (see Figure 5.2). When a system produces products and services with defects and variation, the cost of addressing defects and unacceptable variation is much more expensive to mitigate that if indeed the defects and variation were addressed in design. And

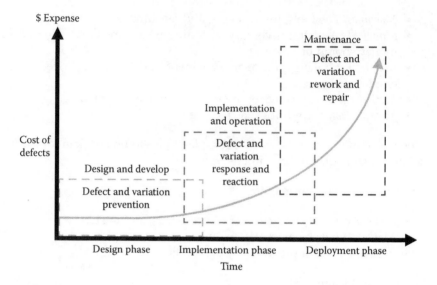

Figure 5.2 Cost of poor quality for a product or service over time.

of course, it follows, when a product or service has been provided to the customer (after deployment), it is of a much greater expense to resolve problems than the cost of addressing such conditions during, or before, the implementation phase. In other words, the cost to resolve goes up as time goes on. The later the issues happen, the more costly they are to fix. The rule is to eliminate the waste of defect and variation and design in quality! (Easier said than done.)

In terms of cost, the cost of quality can be expressed in terms of work (see Figure 5.3) using the people perspective as the definition nucleus. Work, under the definition of affordability, can be expressed in three distinct categories of cost: standard work, conformance and compliance work, nonconformance and noncompliance work. Following the basics in terms of cost of quality originally defined by Philip Crosby, the cost of good quality equals standard work costs plus conformance and compliance costs. The cost of poor quality is made up of nonconformance and noncompliance costs. Standard work, characterized by "best practice," produces good quality at effective and efficient expense. Necessary value-added support work provides the effort of ensuring and assuring quality. All other costs can be categorized within the realm of the cost of poor quality.

The efforts of affordability focus on improving value, increasing customer, and decreasing cost in the Shigeo Shingo, "easier, faster, better, cheaper" fashion. Whether implementing a "just-do-it" (a common sense instant win), a rapid improvement event (a quick win), a project

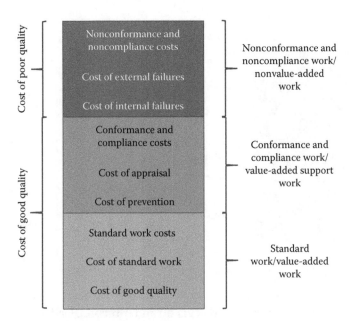

Figure 5.3 Cost of quality in terms of work.

implementation (an operational or system improvement), or a program initiative (a more strategic long-term win), the intent of affordability is to move an organization from its current state to an improved future state, from a beginning to a target designed and planned by the organization. From the successful activities, events, projects, and programs I have been exposed to, the results and achievements are always the same (see Figure 5.4). Unsuccessful efforts are typically focused on only zero, one, or two of the elements of affordability, and not all (refer to the house of affordability). As an example, focusing on cost and cost alone, organizations do not benefit in terms of value and customer, often reducing value and negatively affecting the customer.

Measuring conformance and compliance, as well as cost, provides an initiative performance scorecard that can be universally applied. As applied to a type of customer, market, product, service or process, conformance to requirements, compliance with standards, and cost from an expense and price focus, drives more standard work, better cost, more qualities, and a residual fund for investment in more improvement initiatives, or reward and recognition of achievement, or another customer or market. At the beginning of many improvement programs, due to history of other similar programs of the past 40 years, the people fear for their jobs and employment downsizing. Beginning with a message of affordability and a validated vision of true improvement, such fears and anxieties can

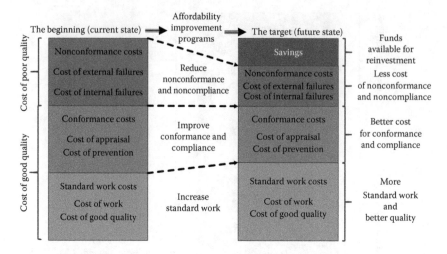

Figure 5.4 Efforts of affordability.

be addressed. What is required is vision and direction, leadership engagement and involvement, the people of the organization, tools and resources
for improvement, and a design and plan that delivers results.

Assuming the path is defined, leadership is fully participating,
and the people are on board, it is necessary to provide effective tools
and resources. One of the tools that is always necessary is a problem-
solving process. Moving forward, it is necessary to resolve and remove
the barriers, roadblocks, and obstacles on the path to success. There are
many problem-solving methods and approaches. One that I use, and
have taught during the 1980s, was a seven-step problem-solving process:
(1) research and understand the problem, (2) identify tools and resources
available to solve the problem, (3) design and plan a solution, (4) build/
prototype/construct the solution, (5) assess/evaluate and validate/verify
the solution fixes the problem (if not, go back to #4 and rebuild and
improve the solution, continue #s 4 and 5 "until the solution is ready"
and approved for implementation), (6) prepare and package the solution for implementation, and (7) implement and maintain the solution.
Strategically, affordability uses (1) assess, (2) design, (3) implement, and
(4) maintain (see Figure 5.5).

Of course, there are other problem-solving processes and methods
that can be used within the pursuit of affordability. One such method,
PDCA, was popularized by Dr. Deming (see Figure 5.6). PDCA, or sometimes utilized as PDSA, is a straightforward approach that can also be
used as CAPD when a process is being measured, or checked, and discovered that a fix or action ("A") is required and a plan ("P") can be created to
do ("D") or perform and improvement.

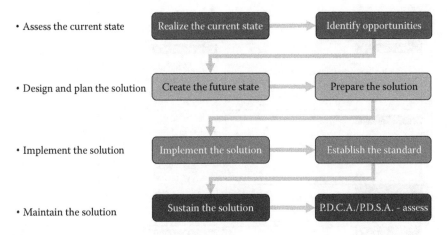

* Assess the current state
* Design and plan the solution
* Implement the solution
* Maintain the solution

Figure 5.5 Assess—design—implement—maintain.

PDCA/PDSA

Act   Plan

Check/study   Do

Figure 5.6 Dr Deming's plan—do—check/study—act method for improvement.

Under the Lean umbrella, A3 is a commonly used problem-solving method (see Figure 5.7). It contains several sections that provide information for planning and designing a solution (background, current situation, goal, analysis). The recommendations section contains the options available to pursue and identifies the recommendation to be pursued as the solution. The plan section documents how the solution will be implemented. The follow-up section articulates how the plan and implementation will be monitored, reviewed, communicated, reported, and adjusted if necessary.

When PDCA and A3 are compared, it is evident that they have a very similar flow and content (see Figure 5.8). It is not surprising since their foundation came from Dr. Deming and the Japanese manufacturing industry.

One other notable problem-solving methodology comes from the scientific method background and has been implemented and practiced

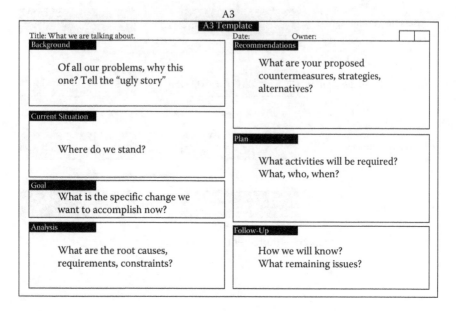

Figure 5.7 The lean A3 problem solving template.

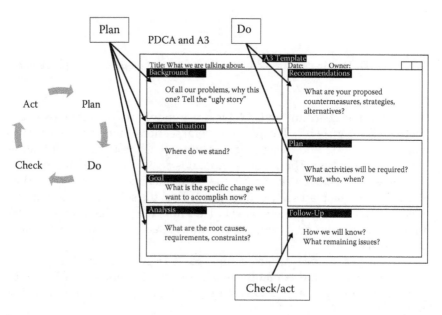

Figure 5.8 The PDCA and A3 alignment.

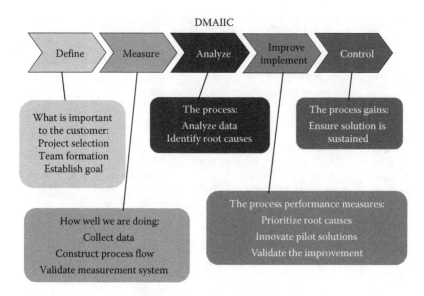

Figure 5.9 Six sigma: a powerful problem-solving methodology.

since the 1980s. As a result of the Six Sigma effort defined by Motorola, and popularized by GE, define-measure-analyze-improve-control (DMAIC) has become quite popular over the past 20 years (see Figure 5.9). This particular version (DMAIIC) has been instituted by the Institute of Industrial Engineers (IIE soon to become the IISE). Since I have served the IIE for several years, I often use DMAIIC. The purpose of the second "I" is to emphasize and reinforce the importance of preparing the solution for implementation and deployment. Each step helps the user (1) understand the problem, (2) obtain the data, (3) study the data, (4) identify and institute a solution, and (5) maintain and sustain the solution. It is important to understand that all of these problem-solving methods, as well as many others not highlighted here, can be used within affordability to make improvements and reduce cost.

Considering Lean and Six Sigma to be two powerful tool boxes for affordability, the tools and tool sets offered are capable of assisting in solving a multitude of problems. Figure 5.10 is an illustration of but a subset of the tools and techniques available for use within affordability. Affordability does not restrict, nor dictate the tool boxes and tools to be used for improving value, better addressing the customer base, and reducing cost, expense, and price.

Of course, it would be amiss to not mention the basic quality tools, the new quality tools, and the entire quality toolbox. Having been a member of the ASQ, I will recommend that the best source I've found for quality

Define	Measure	Analyze	Improve implement	Control
Benchmarking	Confidence intervals	Affinity diagram	DFSS	Gemba/quality walks
FMEA	Measurement system analysis	Brainstorming	DOE	Performance boards
IPO diagram	Nominal group technique	Cause and effect diagram	Kanban	Control charts
Kano's model	Pairwise ranking	Spaghetti diagram	Mistake proofing	Control plan
Knowledge-based management	Physical process flow	E-test	PF/CE/CNX/SOP	Reaction plan
Project charter	Process capability analysis	F-test	Standard work	Run charts
SIPOC model	Process flow diagram	Fault tree analysis	Takt time	Standard operating procedures
Quality function deployment QFD	Process observation	FMEA	Theory of constraints	
Voice of customer	Time value map	Histogram	Total productive maintenance	
House of quality	Value stream mapping	Historical data analysis	Visual management	
Value	Waste analysis	Pareto chart	Work cell design	
Value stream		Reality tree	5S workplace organization	
Flow/pull		Regression analysis	Kaizen/rapid improvement events	
Waste		Scatter diagram	Reliability	
		t-test		
		5 whys		

Figure 5.10 Some lean and six sigma tools viewed together.

tools and the quality toolbox is on the asq.org website. The htmls to locate these tools and the toolbox are as follows:

- *Seven basic quality tools*: http://asq.org/learn-about-quality/seven-basic-quality-tools/overview/overview.html
- *Seven new quality tools*: http://asq.org/learn-about-quality/new-management-planning-tools/overview/overview.html
- *The entire quality toolbox*: http://asq.org/learn-about-quality/quality-tools.html

Quality and performance excellence can be achieved through problem solving using the various tools and toolboxes. Regardless of the problem-solving process you use, the tools and toolboxes are applicable. I've learned in my profession, in order to remain flexible and serve all types of customers and industries, familiarity and agility is key critical. That is why I don't proselytize a single method or approach. When it comes to Lean, Six Sigma, theory of constraints, value-engineering/value-analysis, total quality management, etc., they all qualify for methods to be applied to affordability. The "one-size-fits-all" philosophy does not hold true given the diversity, variety, and assortment of all the organizations seeking more value, more customer, at less cost.

It also holds true that although quality problems can mostly be grouped and categorized using similar characteristics, each specific quality problem for each organization has its own unique traits and dynamics. Like a good mechanic, it is up to the people involved to choose and use the right tools and materials for the job at hand. Within affordability quality, it is the responsibility of the people to make better. In order to make better, the right approach and the right tools and materials, the right information and the right resources are required. I have more than 150 case examples of problem solving to "make better." Below are a couple of case examples of affordability and making better.

Customer quality case example: NCR—retail product total quality

Often, quality is measured by defects of a product or a service. Consider the situation where the product is 100% operational, and the service personnel installing the product are the highest quality service in the industry, but defective conditions arise. Many years ago, working as the director of quality for NCR's retail systems products, I encountered a condition where, of the tens of thousands of sites scheduled for installation of an excellent product installed by exceptional personnel, a very small percentage of the sites experienced a defect.

The products were being delivered at a 100% functional and operational level due to a system run-in that was performed at the plant validating that all of the units ran perfectly, and a validation run-in was also executed at the installation site. It was verified that every unit was functional and operational at the installation site before installation. While the field service personnel installed the systems at a customer site, they experienced, in rare instances, a failure and nonconformance of the system ordered. This condition created a dissatisfied customer. Their expectation was that each and every unit would be installed and operational before opening the retail store the next day. Installations were typically planned for afterhours so that the customer could open and operate their store the next day. Some stores required two or three units (e.g., Lerner New York), while others had as many as 200 units (e.g., J C Penney). Although the problems did not prevent the store from opening, it did impact the satisfaction and approval of the customer. With a production of over 80,000 units, this condition occurred about 100 times, or less than 0.5% of the time. Although the frequency of occurrence of these anomalies was rare and quite minute, even one occurrence in a store chain of over 1000 stores was communicated by the customer to the senior vice president of the division, my boss.

As a response to this condition, a team was formed from plant personnel, service personnel, and "friendly" customer personnel. The first action was to perform a team assessment in order to understand the conditions and the root causes of the customer dissatisfaction. A pareto chart was created from the statistics revealing that 80% of the customer problems were caused by 20% of the various types of nonconformances. The 10 nonconformance types that amounted to 100% of all the nonconformances disclosed that the two major culprits representing 20% of all of the causal factors was missing cables and wrong cables (the defects occurred 50 and 30 times, respectively).

The NCR facility was established and launched in October 1989 as a just-in-time (Lean design assembly plant) for the purpose of producing and delivering NCR retail products worldwide. The system flow map (see Figure 5.11) explains the flow of information and material throughout the plant. The product was built starting at the point of incoming material on the upper left, through the receipt and distribution of customer orders to each manufacturing area, along the path of subassembly and assembly, to order consolidation, packing, and shipping, and finished goods were sent out the door on the upper right of the diagram. The team decided to put a temporary procedure in position 8 dedicated to stopping and resolving every order exhibiting the two primary failure modes. From the time of the assessment (step 1), the team also worked on a design and plan to permanently resolve the issues (step 2). During problem solving, identifying opportunistic solutions, and prototyping fixes, the team invented a method for ensuring the right cable was packed with every order. The fix involved

- On every order, an easy-to-read, color-coded, item request was clearly printed.
- The cables were made available in the pack and ship area in a pigeon hole matrix that was clearly marked by size and color coding for cable type.
- A special location in every packing box was designated for cables.
- The last step in the process was to pack the proper cables and verify by checklist that the cables were placed in the box.

This solution eradicated the nonconformance, and while solving that problem, the team was able to define a solution for the other eight defects.

Lessons learned from the event spawned a number of other organization improvement practices:

- The concept of team-based problem solving was the method of choice for resolving process performance issues.
- The practice of taking a team to an installation site began a program named "field quality audit." It consisted of a team of three employees (called "associates") traveling to an installation site to witness, first hand, the implementation of products created at the plant.

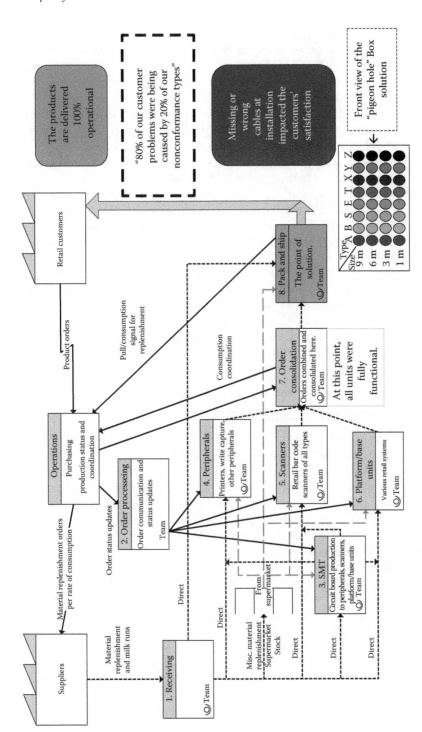

Figure 5.11 NCR retail systems products value stream map.

The team was configured of one manufacturing associate, one engineer, and one other associate (from HR, finance, quality, support, facilities, etc.). The team was commissioned to travel to an installation site, observe, and film the installation, and create a summary report of the result.
- Visuals from videotaping were used to present the actual process execution. Every 3 months (every quarter), the videos were compiled and presented to the workforce at the quarterly meetings.
- Some of the associates were even sent to Sydney, Tokyo, and London as a reward for excellent work at the plant.
- The customers as partners were extended to all customers, and plant visits were scheduled every day (sometimes twice a day) to show the customers how the products they ordered were being made.
- A couple of years later, an employee performance bonus program was instituted that paid associates a portion of their salary based on the system's accomplishments on the key performance metrics and measures: customer satisfaction, employee satisfaction, on-time delivery, quality, and cost/profitability.

From a perspective of team problem solving, a number of additional quality initiatives were employed.

Service quality case example: Anixter–Lucent, supplier–customer relationship

The Anixter–AT&T, supplier–customer relationship was created around the middle of the last century. As the relationship developed, when AT&T created Lucent Technologies in 1996, a relationship between Anixter and Lucent Technologies was formed. As Lucent initiated its deployment of cell towers, its relationship with Anixter was extended to include the distribution of electronic components to each and every cell tower installation. This business partnership was worth about $200,000,000.00 in revenue to Anixter annually.

Although there were several points of distribution, the flagship warehouse for this partnership was located at Alsip, Illinois, just south of Chicago. After several years, Lucent chose to execute a Lucent Supplier Assessment in January 2005 at that site. The results were disappointing. Anixter scored 68 out of a top score of 100 (in other words, Anixter failed). From that point forward, Anixter was bound and determined to rectify that failure and resolve the largest area revealed as critical conditions relating to quality processes and issues. That event spawned a reactive and panicked response within the management ranks that rippled down from the corporate office in Glenview, Illinois, through the general manager, to the warehouse, creating fear and anxiety in the workforce. The timing could not have been worse. A union had approached the workers in an effort to organize.

At the time, I had been working with another Anixter location of another division. I was asked to come to Chicago and meet with Jeff Matz, who was picked to help lead the effort to "turn the place around." After meeting with Jeff, we went in and met with the assistant vice president of the division who was being held accountable along with the plant GM for the overall results by the senior vice president of the division (Jeff was responsible for implementing a solution).

Our first action was to visit the Alsip location and assess the current condition. The building was organized like a typical warehouse (see Figure 5.12). The material flowed in one set of doors at the receiving dock, then it was put away and inventoried, then it waited until an order dropped and a pick ticket was issued, then a "picker" picked the material and delivered it to pack and ship, and finally, it was packed and shipped to the location printed on the order. It seemed quite simple, straightforward, and "normal." Standing there and observing the process first hand, it was obvious, from the level of chaos, that the system was neither stable nor capable. Piles of material appeared in receiving, piles of material in packing and shipping, employees running around "with their hair on fire" (a common phrase they used), and there seemed to be a lot of conflict on the floor. We talked with, and interviewed, several people in each area of the warehouse. It became obvious to me each function was a silo unto themselves. Protecting their own work and

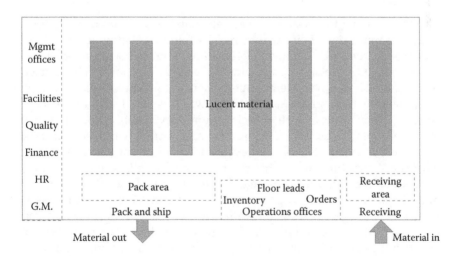

Figure 5.12 Anixter's lucent warehouse, Alsip, Illinois.

blaming other functions for problems and issues. Before we left, Jeff and I went in to the GM's office and met with him for about 15 minutes. It was obvious he intended to communicate he was quite busy and we were getting in the way. In fact, his biggest priority was about the union organization vote coming up and not the performance of the warehouse. To me, it was intuitively obvious the place was "ripe for picking." The GM managed by fear and his priority was not on quality service, but instead on his own survival.

Our assessment validated Lucent's assessment. Leadership, people, process, and performance were all failing. Each function in the organization was operating as a silo unto themselves. Receiving did receiving, and only receiving. Put-away performed the put-away function. Inventory handled inventory. Pick picked product and delivered it to pack and ship. And, pack and ship, packed and shipped material to the Lucent location designated on the order. Each group had their own territory and operated independently from every other function. In fact, Q/A had the responsibility of inspection. Inspection caused a lot of delays in the pack and ship area, and it was no wonder orders were often delivered late.

Jeff and I decided we would address the people first. This was obviously the biggest pain point at that location. Our approach to the people was to introduce the concept of integrated process or warehouse team system using an internal customer/supplier design. By mapping the processes, it was quite clear and easy to introduce a system's approach:

$$\text{Suppliers} \rightarrow \text{receiving} \rightarrow \text{put-away} \rightarrow \text{inventory} \rightarrow$$
$$\text{orders} \rightarrow \text{pick} \rightarrow \text{pack/ship} \rightarrow \text{customer}$$

Each point along the system had a their "suppliers," "input," their "process," "output," and their customers. The traditional and simple SIPOC addressing internal customer relationships. The first activities planned were designed to deliver "quick wins" for each group, now being referred to as team (i.e., receiving team, put-away team, inventory team, etc.). By documenting their process, each team was able to clearly understand their supplier inputs as well as their customer expectation and their own team's outputs. It also fascinated me how each worker was fixed and static with their own job and role, not even interested in understanding their coworkers' functions and what their internal suppliers and customers did for a living. I had to admit that the site was ripe for organization. Jeff and I took on the task of organizing the plant ourselves using process as the core and people as the catalyst. To say the least, this approach threatened the GM's pride and ego. Later that year, he had to be replaced, not as a result of our recommendations, but due to his reluctance and resistance to change. I can truly say, even in this instance, I have never had to recommend downsizing nor firing. I have realized that in the pursuit of affordability the resources will eventually be needed due to the outcome resulting in greater demand, customer base growth, and an increased market share.

Our immediate approach was to pick a problem occurring in each team area, introduce the SIPOC concept, use teamwork to solve the problem, and celebrate the results. Each of the problem-solving events chosen for action lasted no more than a few days, so within the six functional teams in the warehouse, we were able to accumulate a few dozen victories in the first month alone. This created momentum and motivation within the workforce. It also established confidence for every team and team lead. As you might guess, when the union organization vote came up, the people overwhelmingly rejected the union because they had power and control over improving and managing their own work functions. This created stability and much needed confidence.

There was one particular problem that appeared that became our first opportunity to solve a facility-wide problem. The electronics and communications cabinets that were being installed at the Lucent implementation sites cost $34,000.00. They were manufactured in North Carolina by a very reputable cabinet maker. The problem occurred every time a cabinet arrived damaged at an installation site. The agreement between Lucent and Anixter called for a "splitting and absorbing of the cost" (a $17,000.00 loss claim absorbed by each company) between the two organizations and replacement of the cabinet. Obviously, it was a loss of time, quality, and cost. Lucent blamed Anixter, Anixter blamed the cabinet maker, the cabinet maker blamed Anixter, and it was obvious, Anixter was caught in the middle. We organized a problem-solving team to assess, design a solution, implement the solution, and maintain the resolution. Assessing the current state condition, the team walked the path that each cabinet took from the time it was received until the time it was shipped, per order, to a Lucent site. Since this was a cross-functional team from all of the work areas, Q/A was able to participate in auditing and validating, incoming cabinets, inventoried cabinets, and cabinets ready for shipment. No defects found, nor process defects identified (they took numerous photos and videos as proof of conclusion). The team decided to work upstream and visit the cabinet maker in North Carolina. That activity served two purposes: the primary purpose—validating the cabinet maker's quality; and the secondary purpose—visiting a high-quality site and experiencing a quality system at work. The results of the cabinet maker's visit; little or no potential for creating defects. The remaining direction for the team to visit was downstream or from the warehouse to the Lucent installation site. To summarize and encapsulate their finding; onsite, the cabinet handling and installation processes did not exhibit the potential for damage, but the cabinet handling by the transportation carrier did demonstrate potential damage. In fact, on one occasion, the problem-solving team was able to video tape a cabinet being damaged as it was unloaded from the truck. So, after a week and a half of assessment, the root cause of the problem was identified; the logistics company, hired by Lucent, was utilizing low-cost and low-quality carriers to reduce cost and save money, while increasing Lucent's and Anixter's cost of poor quality. The solution, use quality carriers for delivery! A simple solution, more so, a big victory that was an "attention getter" at the corporate office.

For the next 2 months, as the system was stabilized, the processes were standardized, and the team/teamwork approach was sustained, a natural energy drove the organization to greater performance. Performance boards were incorporated to communicate and publicize the advancements being made. The ultimate result: quality and performance improved.

After a few months in Alsip, our next challenge and target, the Memphis location, was designated and declared for focus in the summer of 2005. This warehouse held the largest inventory position for Lucent material and component distribution. When we arrived at the Memphis facility, and began our initial assessment of the current state, it was quite obvious that the organization, the leadership, and the system in place were far advanced beyond the level of performance of the Alsip warehouse. In fact, we were able to capture some best practices for implementation across all of the Anixter Lucent warehouses. We spent a lot of time there documenting processes for deployment and implementation across all locations, as well as helping them prepare for their upcoming Lucent Assessment (the same one that the Alsip location received 6 months prior and failed).

One notable event during that activity occurred in late August that year. Hurricane Katrina was bearing down on the Gulf Coast, seemingly heading for Louisiana and Mississippi. On Friday, August 26, 2005, I was scheduled to return home to Atlanta that afternoon. During the day, governors Kathleen Blanco of Louisiana and Haley Barbour of Mississippi declared states of emergency in their respective states. As they did, as a matter of course for hurricanes threatening the gulf, the Memphis site began loading convoys of trucks prepared in accordance with the threat level and the disaster potential. The equipment was scheduled to arrive just after the hurricane so that emergency responders could re-establish communications in the disaster zone. The Memphis teams spent the weekend completing the tasks, and on Monday morning, August 29, 2005, the fleets of trucks were on their way to Alabama, Mississippi, and Louisiana. When the convoys were met at the edge of the disaster zone by state police, the Alabama and Mississippi trucks were welcomed in, unfortunately, the Louisiana-bound trucks were delayed for a couple of days, and the communication recovery efforts were also delayed.

Measure	Baseline	Outcome
Timeframe	Jan 2005	Dec 2006
Average order quality	67%	95%
Chicago assessment	68 pts 01/05	95 pts 12/05
Memphis assessment	93 pts 11/05	99 pts 03/06
Cost/waste reduction	—	$2MM+
New business agreement	—	$40MM

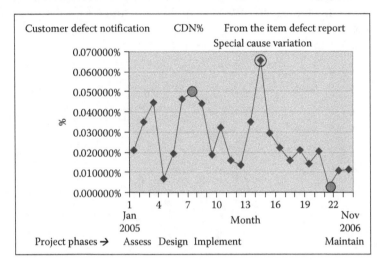

Figure 5.13 Anixter–Lucent component distribution.

The overall effort was a great testimony to the dedication and effort of everyone in that plant.

After 2 years of effort, focused on change and transformation of the Anixter Quality System, using mostly quality tools, Six Sigma techniques, change management, and transformation leadership, the outcome and results were outstanding (see Figure 5.13). The Lucent assessment scores dramatically increased, the quality improved, the costs went down, and the business was continued.

For affordability, there is a recommended path to follow using all the methods, procedures, processes, tools, toolboxes, and techniques available to solve the problem(s) at hand:

1. *Assess*: Assess the current state, understand the problem(s), identify opportunities to improve the process and "make better."
2. *Design*: Design and plan a solution, create the future state, prepare the solution for implementation.

3. *Implement*: Implement the solution, improve the process, establish the standard or best practice that solves the problem
4. *Maintain*: Maintain and sustain the solution, monitor and review the performance status, adjust when necessary or reassess and make better.

The architecture of the house of affordability contains "better" for the sake of quality and capability. An overall intent and effort of affordability is "make better!"

chapter six

Leadership
By any other name is not management

> The world has 6 billion people and counting. We
> need to help 500 million people become better lead-
> ers so that billions can benefit.
>
> **—John P. Kotter**

I've found the best way to understand leadership is to experience it first
hand with individuals who demonstrate the actions, behaviors, attributes, and traits that move organizations from one state of existence
to a higher state of performance and success. The reverse is also true
for learning about bad leadership from those who have taken organizations from success into failure. It is critical to understand both ends of
the leadership spectrum. My first experience with successful leadership
came as a young athlete in high school where my coach (Matt Lococo),
and the team's head coach (Dick Saltrick), took on a losing team and
produced a successful winning program, eventually winning the state
championship.

Over the past 40 years, I've experienced numerous individuals of both
types and many in between. Specifically, within the last 25 years, I've been
exposed to several examples of leadership from the successful perspective that demonstrated the characteristics required to transform organizations. Quite a few books have been written on leadership, but there's
nothing better than undergoing change, transformation, and accomplishment with good leadership.

During the time that I worked for NCR (1980s and 1990s), I passed
through the door of quite a few leaders and managers. The company was
very well managed; however, it had lost quite a bit of its leaders by the
time AT&T purchased it in 1991 after having been in business for over
100 years. Although, historically, the merger was seen as a failure, the
retail division from 1990 through 1994 increased its performance. During
that timeframe, the company was under the guidance of a few different

CEOs, and despite the shifts and changes at the top, a dramatic improvement was realized:

Performance area	1990	1994
□ Associate satisfaction	52.0%	65.0%
□ Customer satisfaction	72.1%	85.6%
□ On-time delivery	60.2%	92.0%
□ Quality/reliability	87.0%	99.2% (*product operation quality = 100%)
□ Profitability	−$3 million	$5 million
□ Equipment revenue	$220 million	$290 million

Using customer focus and a people-centered approach, the leaders of the organization put in place a performance system of both qualitative and quantitative measures that gauged performance through satisfaction (customers and employees) and system results (time, quality, cost, profit, and revenue). Even through the turmoil of the merger and the CEO changes, the leadership team demonstrated proof of concept for affordability leadership.

Affordability leadership, as is affordability, is also based on the integration of customer, value, and cost for continuous improvement. Affordability leadership must create the values, principles, vision, mission, and strategy for setting direction, aligning the resources, motivating the people, communicating the message, and executing the plan. However, in order to do this, the leader and/or leadership team must possess and demonstrate actions and behaviors with specific attributes that develop people, concentrate on the process of work, and increase performance from both a qualitative and a quantitative dimension. Leadership for affordability requires someone who can direct individuals and teams, increase system capability, speed and quality, and, finally, measure and modify accomplishment and achievement according to both human and functional factors. In addition, other aspects are present with leaders of affordability; purpose, vision, values, mission, strategy, systems, structure, alignment, and integration (Figure 6.1).

For the past 25 years, I've been studying leadership and researching the effective foundational elements of the true leaders I've encountered. I have not discovered one leader who has emblazoned all of the 35 attributes I've defined that describe great leadership in affordability. Some of the attributes of affordability leadership are adopted from the gurus (proven by observation and experience with successful leaders), while others were defined through first-hand experience and observation. I have several actual vignettes that describe and define the attributes that I'll detail later in this chapter. But first, I'd like to share a handful of consulting encounters that led me down the path to the definition of affordability leadership.

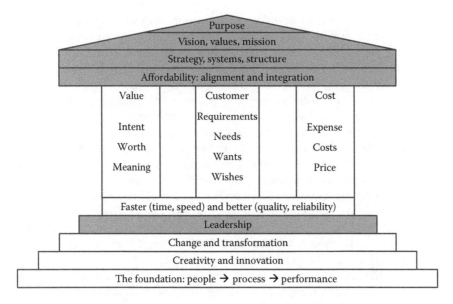

Figure 6.1 The affordability architecture or "the house of affordability."

In the late 1990s, during the first 5 years of my consulting business, I encountered an individual, Jon Hayward, who was cofounder, chief scientist, and system developer for retail transaction systems for Store Automated Systems, Inc. (SASI). Beginning in December of 1996 and lasting until January 2000, I coached and mentored his organization in a variety of areas and organizational excellence. During my tenure there, I was able to observe his approach, engagement, and involvement with the workforce. My first involvement with his organization focused on the technical help desk and service personnel. The second and third comprised engineering and development, and strategic planning. Over the 3-year horizon at SASI, I supported Jon in bringing the company from a $13.5 million value to the $54.0 million level. Because of his attention to his vision, direction, alignment, and motivation of the people, he led a very successful transformation.

During the early part of the twenty-first century, I was engaged with several other projects for Anixter, Inc., a large and very successful electronic components distribution company during the 2005–2006 timeframe. There were two memorable leaders there who, although not VPs and AVPs, led the organization through some very successful challenges with a major client: Lucent Technologies, representing a $200 million a year contract. The focus of this situation was quality and quality service, where they move the "performance needle" from a failing 68% assessment level to 99% performance excellence.

Their attention to people, process improvement, and performance was not only notable, but their direct involvement and contribution was stupendous. Jeffrey (Jeff) Matz, at the time a corporate "troubleshooter," and Robert (Bob) Beck, the Memphis, TN, warehouse manager, collaboratively drove the improvement initiatives and accomplished outstanding results. Since then, they have been promoted and recognized throughout the company.

As an outcome of increasing needs and requirements of the U.S. Navy and Marine Corps resulting from two wars raging in the Middle East (Afghanistan and Iraq), the demand for F/A-18 Super Hornet Fighter Jets was placed upon Boeing and Northrop Grumman (NGC provided 60% of the statement of work for Super Hornet Production). Two NGC leaders, Mr. George Vardoulakas and Mr. David (Dave) Armbruster, and an IIE colleague of mine, Dr. Elizabeth (Beth) Cudney, collaborated in leading a project from 2007 through 2009 to increase the capacity from 42 to 67 units per year, decrease the cycle time from 5.5 days to 4.0 days per plane, and decrease the cost by $5.1 million. By using Lean techniques, calling the program "affordability," and including the entire workforce and suppliers in improvement initiatives, the results of the program culminated on August 10, 2010, when the U.S. Department of Defense ordered 124 units at $49.9 million each (which was $5.1 million below the 2006 price of $55 million). With the corporate leadership of George and Dave, and the consulting leadership of Beth, the outcome culminated in a "job well done."

As a result of the use of deadly IEDs in Afghanistan and Iraq, there were two extraordinary leaders who headed up the U.S. government's Mine Resistant Ambush Protected (MRAP) vehicle program from its inception in October 2006 until its closure in September 2013: Mr. Paul Mann and Mr. Dave Hansen. In a period of 3 years (mid-2007 through mid-2010), more than 27,000 vehicles were produced and fielded that qualified the program as setting the record for the fastest fielding of a ground combat system, going from concept to use, faster than anything seen in history, including WWII. Beyond being a logistics miracle, the maintenance and sustainment efforts were also magnanimous. Their leadership of a group of hundreds of individuals, six OEMs, hundreds of suppliers and contractors, protecting tens of thousands of warfighters was a feat much greater than anything I've ever experienced. Their efforts as leaders more than qualifies them for recognition, acknowledgment, and gratitude.

The previous four examples, with the involvement of eight leaders, is but a snapshot of the examples and leaders I've used to create the definition of affordability leadership.

Background of affordability leadership

Toward the end of the twentieth century, there emerged a great deal of research and information focused on leadership. This continued prolifically into the beginning of the twenty-first century. In addition, at the end of the twentieth century, we also experienced an inordinate amount of emphasis on management. The two are often confused today and blended together in such a way that both are used in the same context and in the same manner to describe the heads of organizations. With the abundance of different models, styles, and designs, there emerged both formidable archetypes and ridiculous prototypes. The works of Peter Drucker, Dr. W. Edwards Deming, Dr. John Kotter, Dr. Rosabeth Kanter, Dr. Stephen Covey, Dr. Robert Greenleaf, Dr. James Kouzes, Dr. Barry Posner, Drs. Heresy and Blanchard, and many others provided a wealth of knowledge and wisdom on leadership. However, there were also those facsimiles invented that served to confound and confuse what leadership is really about.

"Lead from behind"?

In my own honest opinion, and from years of observation and research, "Lead from behind" is the most ignorant and senseless contemporary phrase I've heard. It's sometimes referred to, and related with, servant leadership, and it was coined by Linda Hill and published in the *Wall Street Journal* in May of 2010. She said, "For now and into the coming decade or so, the most effective leaders will lead from behind, not from the front—a phrase I've borrowed from none other than Nelson Mandela." This phrase was concocted from Nelson Mandela's example of a leader shepherding a flock. There are two flaws in her interpretation and approach: (1) people of their own freedom and will do not behave like sheep; and (2) people look to the leader's behavior, actions, and examples for direction, alignment, motivation, communication, and execution. I personally believe Mr. Mandela's intent of using the context and paradigm of a flock of sheep was meant to refer to servant leadership in which the leader serves and supports the needs of the team as a fully integrated participant, and not an individual leading from behind. After all, you don't lead sheep, you herd them. In person-to-person communication, the true message people understand is a result of leadership and the leader's actions and behavior (i.e., 55% of understanding comes from actions and behavior), the vocal pace and inflection of the leader(s) (i.e., 38% of understanding comes from vocal pace and inflection), and the actual words that are spoken (i.e., only 7% of understanding actually comes from the actual words spoken). People, when assimilating a message, watch what the leader does, observe how the leader speaks, and, to a much lesser extent, pay attention to the words. From the groups I've led and the teams I've coached, and with other groups and teams I've observed and studied, when the leader is behind, not fully engaged, demonstrating a distance or detachment, and "telling" them what to do instead of displaying how to do it, mature and experienced teams and individuals see through the façade, become demotivated, and perform at a lower-than-capable level. Often, the immature and less-than-competent teams benefit early in their maturity, but as they mature and develop capability and competency, the leader becomes of no consequence, and the once sought cohesion of the team and leader dissolves. However, on the other hand, if the team is very mature and highly competent, they could operate in a self-directed manner, with coaching and mentoring, and with very little need for a leader to be present full time. Either way, for the most part, teams and people are not well led from behind.

Affordability leadership is about setting direction, aligning resources, motivating people, communicating the message, executing the plan, modeling the way, inspiring and shared vision, challenging the process, enabling others to act, and encouraging the heart. This is the definition of real leadership. Many of the leadership experts that have had influence on the definition of affordability leadership including Dr. John Kotter, Dr. W. Edwards, Ret. Rear Admiral Grace Hopper, Philip Crosby, Peter Drucker, Jack Welch, Dr. Stephen Covey, Dr. James Kouzes, Dr. Barry Posner, Dr. Teresa Amabile, Dr. Rosabeth Kanter, Dr. John Maxwell, Peter Block, and Dr. Robert Greenleaf. One interesting phenomenon surfaces when it comes to a discussion of leadership and management. Back in the early 1980s, I understood that leadership and management were the same and could be used interchangeably. But as my understanding matured through experience, knowledge, and awareness, I was soon to discover that the two are actually quite different. When working with organizations, it's not uncommon to encounter people with varying levels of understanding of leadership and management.

Leadership versus management

One thing is certain, that leadership is leadership and management is not leadership. Quite a few years ago, 1982 to be exact, I was invited to speak at a micro-computing conference in North Charleston, SC, in what is now called the "Charleston Convention Center." The keynote speaker was the well-known Retired Rear Admiral of the Navy, Grace Hopper, also known as the "mother of COBOL" and the one who coined the phrase "computer bug." Her topic was the history of computing, something she had personally experienced throughout the past several decades. One of her key points that stuck with me through the years: "You can manage things, but you can't manage people. People are unmanageable, they must be led." And as I matured as a leader in my own right, her point became clearer and clearer. The things of management are unique, and the people's requirement of leadership is critical to success. Over the past several years, I've witnessed the attempts of managers trying to manage people, resulting in dismal failure. But "managers," leading people, prove to be fruitful. Ms. Hopper was right; people must be led and have a leadership presence in place in order to succeed. Leadership is about change and transformation, management is about "status quo." There exists a natural conflict between those leaders who desire to move the organization forward with those managers who prefer to keep things the same. This dynamic in itself produces much of the challenge when pursuing the improvement of an organization. It requires leadership for improvement. In order to succeed, leaders must integrate management into the vision and strategic direction. Both leadership and management must be in cooperation and integration when change and transformation are required.

It is leadership's responsibility to keep the organization viable through change and transformation. It is management's responsibility to maintain viability through maintaining stability and status quo. It is leadership's responsibility to change and transform strategically, while maintaining the stability and status quo operationally. It takes knowledgeable, capable, and savvy leaders to accomplish such a feat. It takes an affordability leader to accomplish the task and meet the challenge. So where do you find such leaders? What do they look like and what do they do? How do they act? Are they born or made?

Leaders: Are they born or made?

A common question of leadership, and a frequent topic of debate: "Are leaders born or made?" I've been able to observe this phenomenon in children. While coaching numerous sports teams, both boys and girls, encountering over 1000 athletes, I've seen leaders who appeared to be born for leadership, I've seen individuals develop into leaders, and I've also seen leaders rise to a leadership challenge, not being born to lead nor developed to lead. So for me, the answer is yes, and then some. My reality for this nature versus nurture debate is also prevalent in adults. Working within academia, NCR, AT&T, as well as with more than 85 other organizations, I've witnessed the naturals, the groomed and prepared, and the ones who had to step forward and lead when the challenge arose, neither being a natural nor a developed leader. But there is one commonality as it applies to affordability; each successful leader acquired their advanced capabilities through their past, their personal paradigm, and their learnings (Figure 6.2).

The dynamics of how leaders develop and evolve.

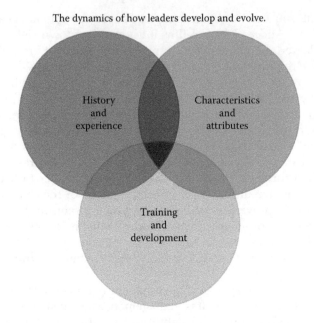

Figure 6.2 Components of the development and evolution of leaders.

History and experience

There was a time in the distant past the great leaders were kings, conquerors, and emperors. For a long period of time, the history and experience of those many generations created a model that taught how to lead. Although today there are many leaders of that same type, the past 200 years of history and experience have surfaced new leadership models that have proven successful. Many of the newer models (world-class leaders—George Washington, Abraham Lincoln, Winston Churchill, Martin Luther King Jr., Nelson Mandela, and Mahatma Gandhi) used contemporary methods of influencing, persuading, motivating, directing, and aligning large groups of people for a common purpose without the use of coercion, intimidation, and oppression. Today, young people have numerous options and choices when it comes to aspiring for positions of leadership. Through their lifetime, depending on the means and method they choose, their history and experience with leadership will be shaped by their current condition as it evolves, events that happen to them over time, and ideas that enter and stay within their mind.

I've been able to observe leaders of all types since the 1950s. Those most memorable, which have stuck in my own mind, have been the best and the worst. Of the worst, I vividly remember those who have murdered and oppressed millions: Fidel Castro, Che Guevara, Mao Zedong (Mao Tse-Tung),

Idi Amin Dada, Kim Jong-Il, Omar al-Bashir, Pol Pot, Ho Chi Minh, Saddam Hussein, Leonid Ilyich Brezhnev, and Robert Mugabe. Of the best, I have witnessed the positive effects of John F. Kennedy, Martin Luther King Jr., Ronald Regan, and Margaret Thatcher. But of course, my personal definition of leadership came from history and experience from those who I have come into contact with most frequently, combined with the examples of those who I have researched and studied most often. Since 1980, I've come in contact with a number of leadership experts and gurus: Ret. RearAdm Grace Hopper, Dr. W. Edwards Deming, Philip Crosby, Dr. Joseph M. Juran, Dr Peter R. Scholtes, Dr. John Kotter, Dr. Teresa Amabile, Robert M. Gates, MajGen John Kelly (USMC), and others who all have improved my understanding of what makes a leader and shaped my definition of affordability leadership. Looking back, I realize they all had that sense of value, customer, and cost in their own unique way. For what they did and how they did it, their value was evident, their customers were obvious, and the cost (sometimes at the ultimate price) was always considered as part of their work and vestige. They characterized and displayed the model of affordability.

Characteristics and attributes

If you research leadership characteristics, attributes, traits, and qualities, you will find quite a few lists that articulate similar personas. Here are four such lists from four different expert groups:

- Honesty, delegation, trustworthy, communication, confidence, commitment, positive attitude, creativity, intuition, inspire, approach
- Vision, courage, integrity, humility, strategic, focused, cooperation
- Honesty, communication, confidence, inspiration, positivity, delegation, commitment, humor, creativity, intuition
- Role model, inspirer, enabler, achiever

After spending several years compiling, integrating, and condensing lists of leadership traits and qualities, I've been able to categorize them into the categories of leadership, organizational, cultural, and personal perspectives, and create a comprehensive list of characteristics and attributes to use on profile leaders of affordability (Figure 6.3).

Each major category has one or two subcategories determining the leader's profile. Leadership actions and behavior are about direction, alignment, motivation, communication, execution, conduct, systems, and empowerment. The organization of people and business dwells with the leader's effectiveness from an institutional viewpoint. The cultural dimension provides a view into the leader's impact on the group's culture. Finally, the qualities that comprise skills and self provide a framework and checklist for the leader's individual personal development.

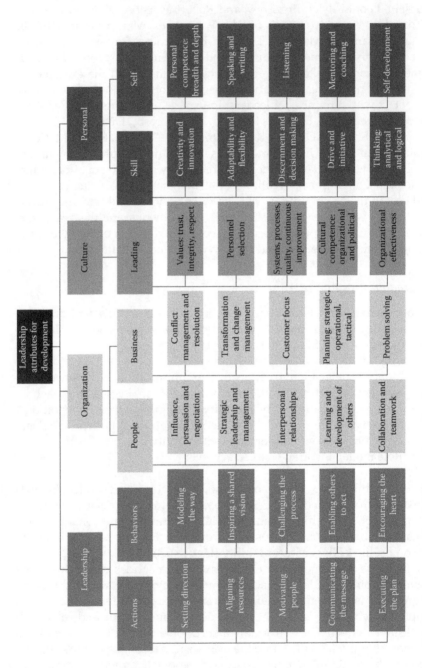

Figure 6.3 The taxonomy of leadership attributes for development.

• Actions	• Behaviors
1. Sets direction	1. Models the way
2. Aligns the resources	2. Inspires a shared vision
3. Motivates the people	3. Challenges the process
4. Communicates the message	4. Enables others to act
5. Executes the plan	5. Encourages the heart

Figure 6.4 Leadership's top 10 attributes.

Of course, I have not yet encountered a leader demonstrating excellence in all 35 areas. However, using the first 10, or as I often refer to them, "the top 10," you can evaluate a leader for affordability effectiveness. You can apply a 1–10 scoring method for each attribute to create an initial assessment for any leader. Over the years, I've used this abbreviated scorecard for assessing a leader from an action and behavioral standpoint. Some leaders can set direction, align resources, and motivate people, but have problems when it comes to communicating the message and executing the plan. Others model the way, inspire a shared vision, and encourage the heart, but fall short when it comes to challenging the process and enabling others to act. The best leaders show quite a bit of all 10 (Figure 6.4).

Training and development

Training and development is critical to every leader's success and ongoing accomplishment. It doesn't matter whether the leader is born or created. The leaders I honor and revere most have and had an appreciation for learning combined with an ongoing dedication to grow and develop their selves. I have also observed leaders that plateau, stop learning, and die. Albert Einstein said, "When you stop learning, you start to die." Tom Clancy said, "Life is about learning; when you stop learning, you die." Henry Ford said, "Anyone who stops learning is old, whether twenty or eighty. Anyone who keeps learning stays young."

I recommend that before anyone jumps into leadership training and development (s)he must follow four simple steps:

1. Assess your current state in terms of leadership.
2. Design and plan a direction and course of action.
3. Implement the solution and follow the design and plan.
4. Maintain the gains and do it all over again ... assess, design, implement, and maintain.

A detailed assessment can be performed using all 35 attributes. With each attribute, using a score of 1–10 (1 = does not exhibit this attribute to 10 = best-in-class at exemplifying this attribute, and anchored with

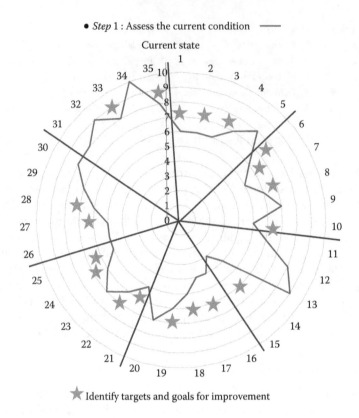

Figure 6.5 Current state assessment.

values of each level in between: 2, 3, 4, 5, 6, 7, 8, 9), leaders being evaluated can be assessed by themselves, an independent party, coaches and mentors, employees, their leaders, or any combination of individuals who may have the ability to assess the leader. The goal is to capture a current state profile of the leader and provide enough information to identify targets and goals for improvement (see Figure 6.5). After mapping the current state and charting the results, the targets and goals can be mapped on the model for providing a path and direction for improvement.

Using the current state map with the targets and goals for improvement, a design and plan for improvement can be developed to serve as a future state vision, direction, and alignment for the leader (see Figure 6.6). After the design and plan is available, the leader should follow the plan, achieve the design, and accomplish the future state, the "Do it!" It is recommended that the implementation be done under the guidance and support of a coach and mentor. The coach and mentor is typically a senior

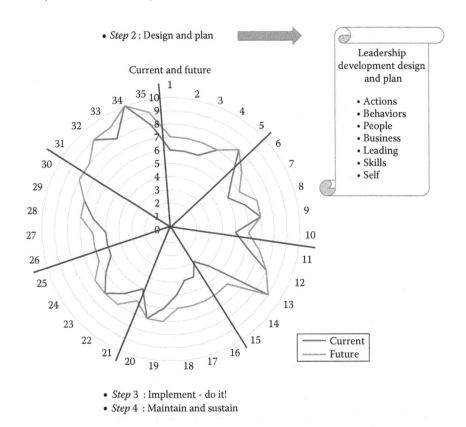

Figure 6.6 Creating the future state.

individual who can coach the leader in the development of skills and mentor the leader such that (s)he keeps the leader on their purpose and in alignment with their direction for improvement.

My recommendation for use of a coach and mentor is purposeful because, although I fervently believe in training, and I frequently participate in delivering training myself, direct personal development is the most effective method when it comes to human improvement. A few years ago, someone sent me a list of 20 points that compare and contrast training and development. The following items point out the limitations and possibilities of training and development:

1. Training blends to a norm—Development occurs beyond the norm.
2. Training focuses on technique/content/curriculum—Development focuses on people.
3. Training tests patience—Development tests courage.
4. Training focuses on the present—Development focuses on the future.

5. Training adheres to standards—Development focuses on maximizing potential.
6. Training is transactional—Development is transformational.
7. Training focuses on maintenance—Development focuses on growth.
8. Training focuses on the role—Development focuses on the person.
9. Training indoctrinates—Development educates.
10. Training maintains status quo—Development catalyzes innovation.
11. Training stifles culture—Development enriches culture.
12. Training encourages compliance—Development emphasizes performance.
13. Training focuses on efficiency—Development focuses on effectiveness.
14. Training focuses on problems—Development focuses on solutions.
15. Training focuses on reporting lines—Development expands influence.
16. Training places people in a box—Development frees them from the box.
17. Training is mechanical—Development is intellectual.
18. Training focuses on the knowns—Development explores the unknowns.
19. Training places people in a comfort zone—Development takes them beyond comfort zones.
20. Training is finite—Development is infinite.

If what you desire is a robotic, static thinker—train them. If you're seeking innovative, critical thinkers—develop them. This pertains to all people of all types and all levels and positions. It is impossible to have an enterprise that is growing and evolving if leadership is not developing and growing along with it.

Of course, my preference, leaders must be developed. Last year, I was in the Republic of Macedonia working with 40 young leaders selected for the Annual International Training for Macedonian Young Leaders program. It's a program that was created by the President of the Republic of Macedonia, Dr. Gjorge Ivanov, during his first term in office 6 years ago. The purpose of the program is to expose young leaders to worldwide success principles in an effort to develop their knowledge and capability for leadership. Over a 2-day period, I covered four topics (leadership, strategic planning, creativity, and innovation) and engaged the group in experiential learning exercises aimed at maturing their leadership skills and capabilities. During the creativity exercise, we harvested a few hundred ideas focused on improving the Republic of Macedonia and the local area known as Lake Ohrid. The ideas were assembled into natural groupings with common themes for use in the innovation exercise. Ohrid and Lake Ohrid, the site for the program, is a tourist destination for Europe, and its economy is based on visitors from primarily European countries.

Although the town is quite clean, the students identified litter along the lake shore as an opportunity for improvement. During the innovative exercise, one of the young leaders identified a "quick win" project of a half-day of clean up for the group to pick up litter along the shoreline. The energy of the group to achieve a project collaboratively was high. The team planned the activity, organized the resources and tools, notified the city's mayor and local television station, and prepared themselves for action. Just prior to the event, the office of the mayor called and canceled the event. The reason: "The local population may become embarrassed by the activity because it is focused on something that is a problem of their community." It is noteworthy to point out, often is the case, I've learned from Dr. Kotter's leadership that great ideas are frequently impeded by fear, delay, confusion, and ridicule, especially in a political environment. The specific learning of this example; if political correctness dictates actions, great ideas are stifled and blocked from implementation, and leaders cannot act. The general learning from such an event: leaders must be permitted to act and exercise what they've learned. Training is important, development is critical.

Of all the organization effectiveness models I've been exposed to, one in particular, The McKinsey 7 S Model, created in the 1970s, has proven to be quite enduring and effective, and very applicable to leadership and organizational training and development (see Figure 6.7). The seven interdependent elements of the model fall into three groupings (i.e., shared

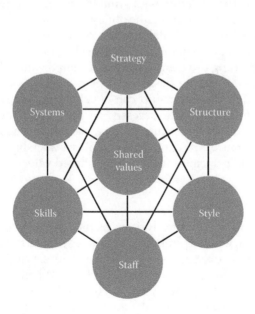

Figure 6.7 McKinsey 7S model.

values, hard elements—strategy, systems and structure and soft elements—skills, staff, style).

The basic premise of the model is that there are seven internal aspects of an organization that need to be aligned if it is to be successful:

- *Shared values*: core values and principles of the organization.
- *Hard elements*:
  - *Strategy*: to carry out the purpose, the way the organization delivers it products and/or services, and the way the organization seeks to enhance its competitive advantage.
  - *Systems*: integrated processes and procedures for delivering products and/or services that contain defined work, measurement, reward, and resource allocation.
  - *Structure*: the architecture of alignment of functions and activities; integration and coordination mechanisms.
- *Soft elements*:
  - *Skills*: the organization's core competencies and distinctive capabilities.
  - *Staff*: organization's human resources, demographic, educational, and attitudinal characteristics.
  - *Style*: typical behavior patterns of behavior of leaders, managers, and all of the people.

Affordability leadership utilizes this model under its originally defined intent for alignment and aligning the organization along with other models (i.e., The Baldrige Framework, Balanced Scorecard) to provide proven reference platforms in the leadership tool box. Leadership can declare, embrace, and operate according to a common bond of shared values. Leadership can define, deploy, and execute design and plan of direction employing the hard elements. Finally, leadership can develop the people by leveraging the soft elements and shape a successful institution that develops, grows, and succeeds with everyone. Affordability leadership is about people, process, and performance. The core of affordability leadership behavior comes from purpose and principles that contains shared values.

Shared values

As for shared values, whether written or unwritten, whether communicated or assumed, whether embraced or supposed, one thing is for certain, organization values exist and shared values endure. Within every organization, values and shared values can be declared. When I worked for NCR while it was owned by AT&T, we indoctrinated AT&T's Common Bond that was known as their code of conduct: respect for individuals, dedication to helping customers, highest standards of integrity, innovation, and teamwork. Affordability leadership's basic three are respect, integrity, and trust. Leaders who demonstrate respect for people, including suppliers, partners, customers, and employees, create an atmosphere that levels the field for instituting integrity and trust. Integrity instills consistency, stability, reliability, and uniformity. Trust is the basis for constancy, enduring interdependence, and teamwork.

Strategy

Strategy is the design, plan, methods, and means by which purpose, shared values, vision, mission, goals, and objectives are accomplished. It also defines how products and/or services are delivered to the customers, how the organization remains competitive, and how the organization will succeed, grow, and improve. The affordability approach to developing and deploying strategy is (1) assess, (2) design, (3) implement, and (4) maintain.

- *Step 1*: Assess calls for the research and understanding of the current state of the organization, the realization of tools and resources required for implementation, the discovery of opportunities for improvement, the definition of performance metrics, measures, goals, and objectives required, and a rendering of the future state image of what the organization will become.
- *Step 2*: Design requires that the diagramming and planning of the strategy is clearly defined, described, portrayed, and explained. The outcome of this phase is a blueprint and plan containing the what, who, where, when, and how to achieve the future state or vision through mission, aim, and purpose. At the end of this stage, the collaborative approach, schedule, and clear message should be readied for implementation.
- *Step 3*: Implement is the period of time when the design and plan is achieved. During this time, adjustments and corrections might be needed due to variations or deviations that may occur. Implementation may last for 1, 3, 5, or more years.
- *Step 4*: Maintain is for the sustainment and maintenance of the design implemented. During this period of time, a new phase of assessment should begin. This approach is designed as a cyclical function that will begin step 1 assess while step 4 maintain is in action.

During the four-step strategic planning process, purpose, values, vision, mission, activities, and actions are defined and deployed. The creation, design, and development of the strategy and strategic plan, containing metrics, measures, goals, and objectives communicate progress and achievement. The people, process, and performance of the system can be monitored and managed through direction, alignment, and pace determined. But, as Dr. Stephen Covey says, "First things first." Step 1 assess.

First and foremost, every assessment should contain a rendering of a current state of the organization. Customers, markets, products, services, competition, positioning, commercial, and financial standing should be researched, understood, defined, and declared. In fact, an ongoing assessment should be maintained on a regular basis for the following:

- *Customers*: Those who purchase the products and services the organization provides.
- *Markets*: The groups of common customers.
- *Product(s) and/or service(s)*: The methods and means by which value is delivered to the customer. Or, the value, produced by the organization, that the customer is willing to pay for and purchase.
- *Competition*: Opposing organization that provide similar value to the targeted customers and market.

SWOT analysis

The process of assess, design, implement, and maintain also requires a comprehensive understanding of the organization's strengths, weaknesses, opportunities, and threats. A complete SWOT analysis should be performed to gain a complete understanding of where the organization

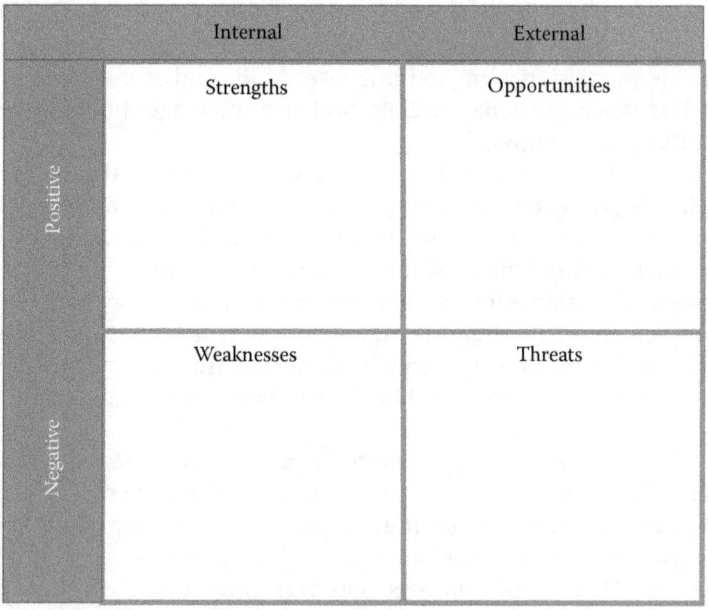

	Internal	External
Positive	Strengths	Opportunities
Negative	Weaknesses	Threats

Figure 6.8 Organization SWOT matrix.

is strong, where the organization is weak, where opportunities exist that the organization can pursue, and where threats reside that may negatively affect the organization. A SWOT analysis template may be used for research and discovery (see Figure 6.8).

Assess the organization in the four SWOT areas:

- *Strengths*: characteristics of advantage as compared to others.
- *Weaknesses*: characteristics that project disadvantage relative to others.
- *Opportunities*: situations or chances that could be exploited.
- *Threats*: conditions that could cause problems for the organization.

The SWOT assessment should serve as research and understanding input for creating or shaping a strategy and developing a strategic plan.

PESTLED analysis

PESTLED analysis is used for assessing and understanding the business environment in a more detailed level, and identifying conditions that have both positive impacts and/or negative connotations. It can be used for pursuing new business, entering new locations, or reviewing the current state in places already established. Each area provides leaders with a strategic view into current conditions that impact designing decisions, planning decisions, and implementation decisions. Knowing the PESTLED details, leaders can evaluate risk, avoid uncertainty, and increase the probability of success (Figure 6.9). The definition and purpose of each PESTLED element is as follows:

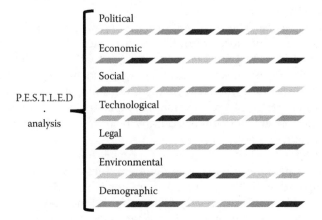

Figure 6.9 Elements of the PESTELED analysis.

1. *Political*: These factors determine the extent to which a government, or even certain political constructs, may influence the economy, or a certain industry, or even a particular product or service. Political factors include tax policies, fiscal policy, trade tariffs, and other impacts that a government may invoke that has an effect on the business environment.

2. *Economic*: These are the factors that may impact a company on both short and long term on the basis of the economy's performance. An inflationary condition affects price and valuation of products and services. Inflation also affects purchasing power subsequently altering demand/supply models. Economic factors include inflation rate, interest rates, foreign exchange rates, economic growth patterns, and economic strength/weakness.

3. *Social*: These factors scrutinize the social environment of the market, and gauge determinants like cultural trends and population behaviors, as well as beliefs, values, and principles. In Western countries, like the United States, buying trends exhibit high demand during the holiday seasons.

4. *Technological*: These factors pertain to innovations in technology that may affect the operations of the industry and the market favorably or unfavorably. This refers to automation, research, and development and the amount of technological awareness that a market possesses.

5. *Legal*: These factors have both external and internal dimensions. There are certain laws that affect the business while there are certain policies that companies maintain for themselves. An example would be in Saudi Arabia where gender laws may conflict with gender policies of a U.S. company. Legal analysis takes into account both perspectives and then considers all possibilities and strategies given both situations.

6. *Environmental*: These factors include all those that influence or are determined by the surrounding environment. This aspect is crucial for certain industries particularly those that may be involved with tourism, farming, agriculture, energy, and mining. Factors of the environmental dimension include, but are not limited to, climate, weather, geographical location, global changes in climate, environmental offsets, and impact on any resources having an impact on that country and region.

7. *Demographic*: These factors are crucial for consideration within today's global economy. The demographics of the United States, as compared with the UAE, Germany, or Ghana display vast differences and contrasts. Specific factors of this kind may produce greater opportunities or even avoid disasters as a result of comparing success and failure aspects under certain demographic conditions.

In some cases, ethical analysis is also included when an organization has the potential of conflict on an ethical basis. Ethics sometimes come into play in conditions where capitalism collides with socialism, one religion may conflict with another, or even when disputes between two sides on gender or racial or legal conditions, practices, and behaviors emerge.

After a PESTLED assessment is completed, the designing and planning activities for solution implementation should take place. However, there may be other dimensions to consider in more detail and include the following:

- Purpose and direction (well-defined and articulate)
- Values and ethics (principled and durable)
- Culture and success (strong sense of identity with the entity and meritocracy)
- Partnerships and relationships (suppliers and customers)
- Adaptability and flexibility (innovation and experimentation)
- Community and environment (participation and harmony)
- Learning and growth (static and dynamic)
- Organization and governance (clear and tolerant)
- Financials and prosperity (conservative and frugal)
- Performance and achievement (accurate and authentic)

Strategic plan design

For strategic planning documentation purposes, any comprehensive and thorough format or framework can be used. I've found that the outline and flow for any specific organization tends to differ due to the variety of people and opinions involved. The specific layout is not as critical as the content. However, I use the 10-point plan (see Figure 6.10) skeleton as a checklist that covers all of the necessary aspects. In fact, for some aspects (e.g., customers and markets targets, products, and services), there may be even more detailed design and plans created.

- Purpose, values, vision, mission, goals, objectives, direction
- Customer(s) and market(s) targets
- Products and services
- Competition and competitive analysis
- Positioning: organization, products, services
- Enterprise Alignment
- Structure: organization and people
- Systems: processes, procedures, methods, measurement, communication
- Plan and performance: measurement, analysis, communication
- Profitability: social, commercial, financial

Figure 6.10 Strategic framework: A 10 point plan.

After undertaking the establishment of shared values and strategy, both systems and structure must be considered. My rule of thumb on which one comes first is found in a common quote of architects and industrial designers: form follows function. Or in this case, structure follows systems.

Systems

Under affordability, systems exist at the enterprise level, and the process level, delivering value to the customer (products and/or services) at a reasonable cost (i.e., organization expense and competitive/acceptable customer price). The system construct for affordability can be understood from Figure 6.11. The enterprise comprises customers, value-added functions, value-added support functions, supplier and partner functions, and customer requirements that are fulfilled and delivered as an output of linked processes that satisfy those requirements by delivering products and services that the customer purchases. This systemic approach should be used as the design center for architecting any affordability organization structure. Before creating the form of an organization, the functionality, processes, and enterprise should be well defined.

Brief definitions of the construct components are as follows:

- *Customers*: Those who purchase the value produced and delivered
- *Value-added functions*: That which the customer is willing to pay for and purchase
- *Value-added support functions*: The functions that support the value-added functions

- Customers
- Value-added functions (deliver products and services)
- Value-added support functions (support value-added functions)
- Supplier and partner functions (support value-added functions)
- Customer requirements (drive products and services)

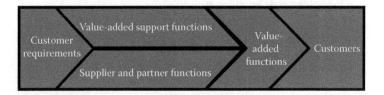

Figure 6.11 Affordability organization system construct.

- *Supplier and partner functions*: External support functions for the value-added functions
- *Customer requirements*: Customer-defined needs, wants, and wishes that describe the value that should be delivered to the customer at a competitive price

Following the functions, the form can then be set.

Structure

Traditionally, and even today, organizational structures are often depicted as two-dimensional charts that show the reporting, governance, and control design and configuration of the institution. Such illustrations are the first step in communicating and conveying an existence of division and are often referred to as "functional silos of operation." Affordability organizations are enterprises and systems that deliver products and/or services to customers. They consider value, customer, and cost as their pillars of strength. Affordability organization structures should be arranged in accordance with how the organization delivers value to customers while managing the fiduciary responsibilities from expense, pricing, revenue, investment, and profitability perspectives (see Figure 6.11). Regardless of how the structure actually looks on paper, the elements of affordability must be present and active.

One major design center for affordability structural design is the use and utilization of teams. Contemporary organizations that have demonstrated the most achievement in terms of affordability use teams and teamwork as the way to get the work done. Although it is often the case that individual heroics are heralded and recognized, collaboration and teamwork are the sustaining facets that demonstrate enduring achievement. Team excellence can be assessed and verified using the following checklist:

1. Interdependence and camaraderie
2. Challenging and stretching of tasks and work
3. Alignment with purpose and function
4. Common language
5. Values of respect, integrity, and trust
6. Shared and dynamic leadership and followership
7. Problem solving
8. Confrontation/conflict handling
9. Assessment and action
10. Celebration

When present, teamwork is at its best. When absent, teamwork is less than collaborative and cooperative. So, within the structure of teamwork, the question of size and span of control surfaces. Affordability supports the implementation of teams for structure and the size and span of control is set at 12. Teams of 5–8 are effective, as are some teams of size 14–18. Teams of 20 or more seem to be too large and lethargic most of the time. However, setting team sizes using "the rule-of-12" is the best de facto method for sizing groups within any given structure. The rule-of-12 for sizing basically states that a leader's team should be around size 12. From observing teams for more than 40 years and integrating the research from the past 20 years, teams of size 8–12 are usually the best optimized size. Some teams, for the sake of specific responsibility and accountability, may be of size 5–8, while others may be in the range of 12–15. Effectiveness begins to decrease around team size 15 and definitely degrades after size 18. It's no wonder that the most popular team sports around the world field teams of 5–15. More seasoned and capable leaders can coach and mentor larger teams of size 12–15, while new or junior leaders may be better assigned to teams of smaller sizes of around 5–8.

Using the rule-of-12, organization layers and levels can be estimated (refer to Figure 6.12). For small organizations, only one layer is required. Sometimes affectionately referred to as "tribal size" and range in size from 0 to 15. After that first threshold, levels or layers in the organization begin to naturally appear. Much criticism of today's organizations have been pointed toward "too many layers in the organization with communication lines being stretched way too thin." In the range of 15–241, only two levels are truly required with the "sweet spot size" being around 157. Three layers would only be required for organizations in the range of 241–3391 with a sweet spot of 1741. Considering this framework proposition, even the top 10 employers in the world would need no more than six levels or structural layers (see Figure 6.13). So, it is safe to say that if your organization has several layers, it is most probably too deep, and not aligned with affordability.

Of course, the underpinnings of affordability are founded in improvement. The affordability design for leading organizational change and transformation is illustrated in Figure 6.14. After having been exposed to both successful and unsuccessful change and transformation efforts, I have found that using Dr. John Kotter's model, employing a parallel operating approach, work best for affordability efforts. My experience with NCR, SASI, Northrop Grumman, and others for the past 25 years clearly demonstrates that this methodology works. It permits the static organization to continue operating at its current success levels while discovery of improvement can be realized and folded back in to the organization over time, thus advancing it toward the aim and the goals.

The affordability leadership definition and model has been created over the past 40 years, from exposure to, and experience with, numerous organization leaders combined with evolving models and examples

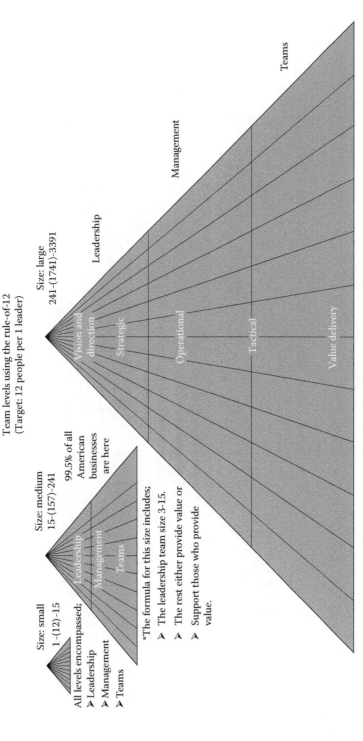

Figure 6.12 Organization structure using the "Rule of 12" for small, medium and large organizations.

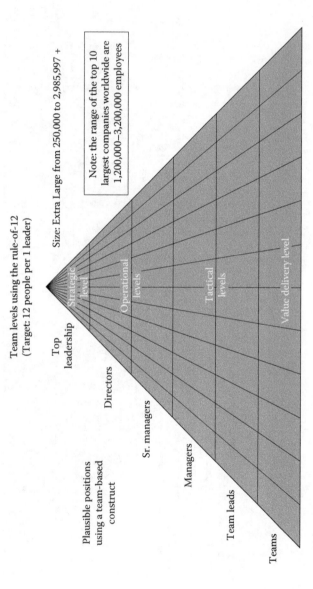

Team levels using the rule-of-12
(Target: 12 people per 1 leader)

Size: Extra Large from 250,000 to 2,985,997 +

Note: the range of the top 10 largest companies worldwide are 1,200,000–3,200,000 employees

Top leadership

Strategic level

Operational levels

Tactical levels

Value delivery level

Plausible positions using a team-based construct

Directors

Sr. managers

Managers

Team leads

Teams

Figure 6.13 The "Rule of 12" as applied to very large organizations.

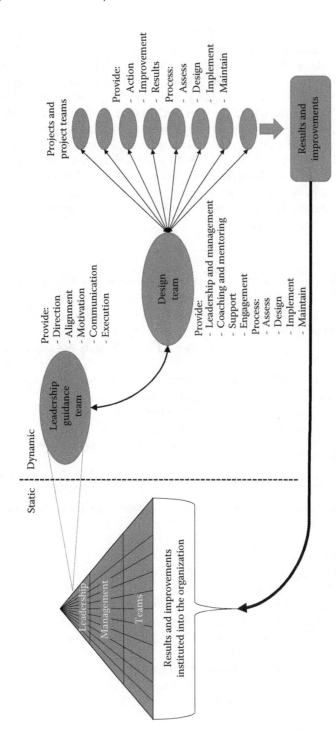

Figure 6.14 Leading teams for change and transformation. (Adapted from "Accelerate," Dr. John Kotter.)

that have demonstrated accomplishment and success. Some of the more recent influencers I've encountered are not military generals, not presidents of countries, not celebrities nor rock stars, but in my opinion, gurus of a special kind. They are extraordinary people who have demonstrated the ability to lead, direct, align, and influence people to do great things. They come from different backgrounds, different environments, different disciplines, with a variety of accomplishments, that all share the similar characteristics and traits of affordability leadership. Since I started my business in 1995, from all the leaders I have encountered, I've compiled a representative list of affordability leaders from 10 of the 85 organizations I served who have had influence, input, and involvement in shaping the meaning and definition of affordability leadership (they appear in chronological order from 1995 to 2015):

Organization	Leader(s)
Harvard Business School	Dr. Teresa Amabile and Dr. John Kotter
Gwinnett County Public Schools	Mr. Alvin Wilbanks
Gwinnett County Tax Commission	Ms. Katherine (Sherrington) Meyer
Store Automated Systems	Mr. Jon Hayward
Anixter	Mr. Jeffrey (Jeff) Matz and Mr. Robert (Bob) Beck
Institute of Industrial Engineers	Mr. Larry Aft and Mr. Don Greene
Shepherd Center	Ms. Susan Bowen
Cantel/Medivators	Mr. David Bazinet
Northrop Grumman	Mr. Dave Armbruster, Mr. George Vardoulakis, and Dr. Elizabeth (Beth) Cudney
The MRAP Program	Mr. Paul Mann and Mr. Dave Hansen

What follows are the "affordability similarities" of these leaders, I've observed:

- Sets direction
- Establishes purpose
- Provides a vision, values, mission
- Executes thinking and action—strategy, systems, structure
- Architects resource alignment and functional integration
- Creates affordability focus: customer, value, costs
- Emphasizes faster and better
- Institutes leadership, utilizes management
- Accomplishes change and transformation
- Enables creativity and innovation

- Grows, develops, and motivates people
- Manages and improves process
- Measures and improves performance
- Accomplishes profitability

Leadership is about people, while management is about things. The balance of both are necessary. However, in pursuit of success, keep in mind that leadership comes before management.

chapter seven

Change and transformation

> It is not necessary to change. Survival is not mandatory.
>
> **—W. Edwards Deming**

Change and transformation is the middle layer or "center step" in the foundation of the house of affordability architecture (Figure 7.1). It is sometimes the most difficult step in the structure to accomplish because it requires alteration, modification, and adaption of the people, the culture, the organization, and the systems involved. Any prescribed or obligatory variation of a static establishment affects the people's perception of stability and constancy. Often, resistance, confusion, and uncertainty occur. It is not surprising to me that more than 70% of change and transformation efforts fail. In order to understand it, let's start with the basic building blocks and definitions.

Change defined

As a verb:

- Make or become different ... "a proposal to change the law."
- Take or use another instead of ... "they decided to change their name."

As a noun

- The act or instance of making or becoming different ... "the industrial revolutionary change from bench work to mass production and assembly line."

Transformation defined

As a noun

- A thorough or dramatic change in form, appearance, or operation ... "the landscape has undergone a radical transformation and it is no longer today, what it was in the past."

Even from a definitive and basic perspective, change and transformation are not exactly the same. The change versus transformation discussion,

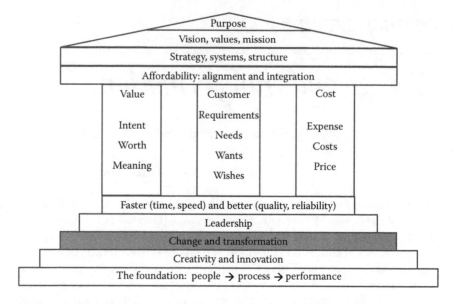

Figure 7.1 The affordability architecture or "the house of affordability."

and confusion, has been going on in management, leadership, and organization effectiveness circles for more than 25 years. When I was exposed to the Organization Effectiveness Group of AT&T's Bell Labs, there was no consensus about the difference between change and transformation. The way I've learned to describe it; change is related to management and managing, while transformation is more about leadership and leading.

According to Ron Ashkenas, in his *Harvard Business Review* article, "We still don't know the difference between change and transformation" (HBR, January 15, 2015), "Change management means implementing finite initiatives, which may or may not cut across the organization. The focus is on executing a well-defined shift in the way things work. It's not easy, but we do know a lot more today about what to do than we did. ... Transformation is another animal altogether. Unlike change management, it doesn't focus on a few discrete, well-defined shifts, but rather on a portfolio of initiatives, which are interdependent or intersecting. More importantly, the overall goal of transformation is not just to execute a defined change—but to reinvent the organization and discover a new or revised business model based on a vision for the future. It's much more unpredictable, iterative, and experimental. It entails much higher risk. And even if successful change management leads to the execution of certain initiatives within the transformation portfolio, the overall transformation could still fail."

Affordability defines change as finite improvements that affect people, process and performance over time, and are typically implemented within operational and tactical realms. Finite improvements in a process, operations, tools, methods, and tactics are all examples of change. Changes occur within groups, teams, and departments and may affect one or many. Change is a natural component within the context of affordability since the aim is to continually increase value, improve customer offerings, and lower costs. The tools and techniques for change can be found within the numerous toolboxes that support affordability (e.g., TQM, TOC, VE/VA, Lean, Six Sigma). In addition, resources for change should be included throughout every change effort (e.g., training, models, manuals, texts, subject matter experts). As a part of affordability, change is an ever constant focused on the triple aim. The standard approach to change should be indoctrinated within every organization.

Affordability transformation, on the other hand, is defined as a complex enhancement affecting an organization and culture that's implemented across all boundaries effecting established paradigms and overall strategy. The tools and techniques for transformation can be found in the numerous strategic toolboxes that support affordability (e.g., re-engineering, balanced scorecard, Baldrige, Kotter's eight steps). Often, transformation tools can be found under the heading of change. It is worth noting that transformation requires change, but change does not require transformation.

Whether it is change or transformation, I have found that resistance is one factor that universally occurs in both, as well as advocacy, and additionally stability and constancy. One might question how all three can be present, and also be curious as to in what proportion. Years ago, Price Pritchett, from observation and research, created the 20–50–30 rule (see Figure 7.2). He declared that there are typically three types of individuals when it comes to change and transformation and they appear

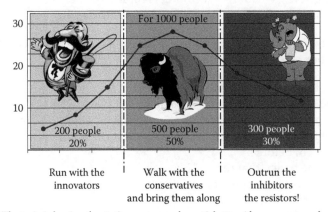

Figure 7.2 "The 20-30-50 rule" Price Pritchett.

in proportions of 20%, 50%, and 30% (note: not always in exactly those quantities but usually close to those percentages). Whenever a change is at hand, this formula can be used and utilized to address the change.

The first category contain the innovators who are willing to serve as champions, advocates, and promoters. They

- Are stimulated by adventure to a degree that they will face the risk of the unknown.
- Are discontented to the point that any change is preferable to the continuance of the status quo.
- Advocate change and are greatly influenced by their urge for change.
- Are collectively the yeast in the "business cauldron."
- Can be mighty irritating because, unlike resisters/inhibitors, they are generally an aggressive, insistent lot.
- Have a fierce devotion to their "child."
- At their best, are refreshing and formidable sources of new strength for the organization.
- At their worst, fanatics with an eye on the grandstand rather than the ball.
- Badge reads, "You can't stop progress."

The second grouping is the Conservatives, the stable, those comfortable with the now, the ones who want to know the right way to go. They

- Respond to ideas in an objective, constructive, and friendly manner.
- Prefer a settled, stable, and predictable life.
- Insist upon results before change, and logical reasoning, by itself, will not galvanize them into action.
- Change only when they see others reaping the benefits of change.
- Definitely wear the badge that reads, "Show me."

Finally, there are the inhibitors, the resistors, the ones who do not want change to happen. They

1. Have ties to the past based on dogma.
2. Oppose new ideas because they are alien.
3. Believe no one can affect significant change since we are all experienced, competent people who have not been able to change, nor transform before.
4. Wear the badge that reads, "No way."
5. Display different types of resistance:
   a. Overt—30%:
      i. Sabotage
      ii. Vocal opposition
      iii. Agitating others

    b. Covert—70%:
        i. Reducing output
       ii. Withholding information
      iii. Stalling (asking for more data or information)
      iv. Appointing task forces and committees
6. Can exhibit conscious or unconscious resistance
7. Can be active or passive
8. Can be well-intentioned or subversive
9. Use at least four well-known strategies for resisting:
    a. Fear mongering
    b. Delay
    c. Confusion
    d. Ridicule (even to the extent of character assassination)

There is a strategic approach to the innovators, conservatives, and inhibitors, those individuals in the various categories of response to change

- Focus on the innovators who drive the change.
- Sell the innovators with logic.
- Sell the conservatives with results.
- Balance the impulsiveness of the innovators with input from the conservatives.
- Focus on the inhibitors last….. if at all
  - Logic or results are not enough.
  - Inhibitors only begin to think about change when they find themselves conspicuous because everyone else has changed. Attention only reinforces their problematic behavior.
- Use the challenging statement.
  - The train is leaving the station, everybody has a ticket to ride. You can ride on the train, you can even take a part in driving the train, or you may stay on the platform as the train pulls away. Either way, we're moving forward. We're going to make progress. Are you on board?
- After you gain some momentum with a few quick wins and some big victories, you could always challenge the resistors directly and ask them
  - "We've made good progress already, and if this is not the right way to go, or it is not good enough, show me a better way! Get involved!"

Additional strategies and tactics include

- Expect resistance. Prepare for it.
- Create allies. Relationships are crucial.
- Make the case for change. Have a strong message.

- Choose your opening moves carefully and make them bold. Quick wins!
- Articulate the destination; overcommunicate. Vision and communication!
- Resolve the "me" issues. The what's in it for me (W.I.I.F.M.) for everyone.
- Involve people to increase commitment. Engagement and involvement.
- Utilize resources effectively. Best people, best skills, best opportunities.
- Promise problems. Engage them is problem solving.
- Overcommunicate. It has been proven, when leaders communicate, 55% assimilation comes from what they do and how they do it, 38% comes from how they say it, only 7% comes from the actual words they say. Be, act, say ... consistently.
- Beware of bureaucracy. Don't build in layers.
- Get resistance out in the open. Conflict is bound to happen, address it.
- Give people the know-how needed. Coach, mentor, and train skills and capability.
- Role model and embody the change values. Walk the talk with integrity.
- Measure results. Measure, monitor, act, and improve.
- Reward change. And reward is beyond just money ... it is also a heart-felt "job well done!"
- Outrun the resisters ... let them leave if necessary, they will continue to hold you back. In fact, volunteer to help them find a better place to be.

The people are about change, and although change is finite, and it takes a lot of change to accomplish a transformation, it must be understood that there exists at least one process for accomplishing change and transformation. The method and approach I embrace strongest has been published and distributed in the many books written by Dr. John Kotter. His well-renowned book *Leading Change* (Harvard Business School Press, 1996) was my first exposure to a process and blueprint for leading change and transformation. Since that time, he has published several texts on the subject from a variety of perspectives, and I've had the privilege to have met him and used his approach. I can testify if it works. The transformation takes eight steps (see Figure 7.3). First, there needs to be a purpose and reason, sometimes driven from a crisis, sometimes from a dream, or even sometimes from both conditions. Next, the utilization of a strong leadership team is critical for success. Subsequently, that leadership team coalesces around a collaborative vision with a design, plan, purpose, value, and strategy known. The communication of the vision and the accompanying

Five principles implied (2014)

- Many change agents.
  Not just a few appointees.

8 _____

7 | Institutionalize new approaches: standardize

6 | Consolidate improvements, produce more change

5 | Plan for and create short-term wins: quick wins

4 | Empower others to act on the vision: empowerment

3 | Communicate the vision: communication, execution

2 | Create a vision: design, plan, purpose, value, strategy

1 | Form a powerful guiding coalition: leadership team

| Establish a sense of urgency: crisis, dream, events, leader

- A want-to, and a get-to.
  Not just a have-to mindset.

- Head and heart.
  Not just head.

- Much more leadership.
  Not just more management.

- Two systems.
  One organization.

"Leading Change," Dr. John Kotter

Figure 7.3 Dr. John Kotter's change process.

message, with the design and plan begins the initiative. Developing and empowering people to act on the direction, alignment, and execution plan for the vision readies the people for action. Although he calls them short-term wins, I refer to them as quick wins designed and intended to create motivation and momentum. With victories and accomplishments realized, more change can be produced and the new approach achieved can be instituted and indoctrinated as standard behaviors and practices in the culture.

More recently, a little more than a year ago, he offered five principles that accompany the eight steps (see Figure 7.3). Many advocates must be employed and engaged. People must have the mindset they "want-to" and need to "get-to" the future state, not just that they "have-to." The hands and the head are important, but it's the heart and the soul that accomplishes transformation. Leadership, leading people is much more important, than management and managing things. And, most successful environments that accomplish transformation operate as two systems. One, the status quo, continues along the path of standard operations and results to keep the lights on and the revenue flowing. The other, a more dynamic system, plans for, achieves, and determines the changes that work.

In using this phased approach for the past 20 years, being involved in change and transformation, I add one additional discovery I've had from my work with AT&T and Bell Labs back in the early 1990s. The American archetype was published in a book written by Lew Hatala and Marilyn Zuckerman titled *Incredibly American*. Although the eight steps appear to be static and constant, my personal experience is that they are stages rather than fixed, sequential steps. I have found the path to trans-formation also involves a more variable and fluctuating "rollercoaster

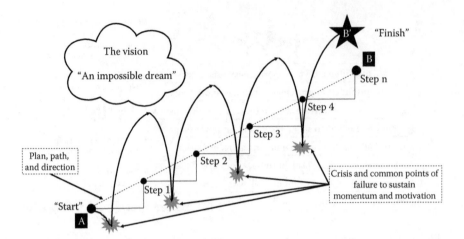

Figure 7.4 The American archetype: "our normal path." (From Incredibly American," Lew Hatala and Marilyn Zuckerman.)

ride" when it came to an organization's motivation, complacency, energy, and emotional response. When it comes to organization transformation, "our normal path" is not a straight line (see Figure 7.4). You would think that if we are at point A and we wanted to go to point B the shortest distance is a straight line from A to B. Or, going from A to B we could take the journey one step at a time. But nooooooo, our path actually starts off at A, in any direction other than toward B, and it takes an event (either a crisis or an impossible dream) to rally our emotional energy to make progress. After some time, we fall into complacency and it requires another "booster" or event to stimulate the emotional energy to make the next advancement. This continues, and often is that case (as it appears in the illustration), we get to B′ before we planned to get to B and B′ is better than what we expected in terms of B. The critical points of failure are likely to occur at those lowest moments if indeed we forget to step in and stimulate the organization's emotional energy to strive ahead.

The three models addressed earlier are the primary strategic paradigms I use for change and transformation. Figure 7.2 focuses on the individual. Figure 7.3 is describes the step necessary for transformation. Figure 7.4 illustrates the path to expect and the basic point along the journey where leadership is necessary and motivation is required. These three are basic tenets of affordability in terms of how to accomplish a transformation. There exists a plethora of theories, doctrines, precepts, and ideologies for transformation. I have found these three to serve me best when it comes to incorporating change and instituting transformation.

The basic building blocks of change and transformation within affordability are

- Purpose, vision, mission
- Leadership
- People
- Processes and resources
- Design, plan

There is a simple formula: purpose/vision/mission + leadership + people + processes/resources + design/plan = change and transformation with increased performance and results. However, every element must be satisfied. Without purpose/vision/mission, chaos and confusion is likely to occur since there is no direction, alignment, and motivation. Without leadership, fear and anxiety arise within the ranks because of the absence of direct support from the leaders. Without people on board, it is obvious that slow or no change is expected. Without processes/resources, aggravation and frustration are liable to appear because there is no "how to" nor any "with what." Finally, without a design/plan, false starts are probable due to failures and setbacks. All five constants of the formula must be present for the change and transformation to occur.

Affordability transformation is for increasing the satisfaction of customers, the people, the suppliers, and partners. It also expects improvement of value, speed, quality, and profitability (due to lower costs, expenses, and prices).

Within the affordability formula, there are numerous roles and responsibilities for every type of person involved. The role of leadership, of course, is to lead the people and lead the change and transformation effort. The role of advocates (people who want the change) is to engage and get involved directly in the change effort. There are targets (people who must do the changing) who must engage in performing the change process, using the change formula and creating change as part of the prescribed activities and actions. The sponsors (senior managers with authority and want the change to happen) must break down barriers and roadblocks inhibiting change. There are change-agents (people who interact with the targets and enable the skills and purpose) to serve the change effort. There are external influencers (people who have influence on any or all of the other four roles) who should support, encourage, and motivate. Then there are the people (those involved in the change) that must have all the pieces in place for them to embrace, enact, and enjoy the change. There is a process for managing change, there is a process for leading change, there is a formula for calculating change, and there is a framework for incorporating change. Affordability is a theory that has proven models and methods for change and transformation.

A key component is communication. During my time with Bell Labs, I was involved in a research project studying communication and what it takes to truly ensure that "the message" was heard, understood, and assimilated. Our finding and proposal was tagged; "the five times rule":

1. The message must be communicated at least five times to be heard and understood by the people.
2. The message must be delivered at least five times using different learning styles: intellectual, visual, audio, experiential, logical, aural, verbal physical, etc.
3. The message must be delivered using at least five different methods: leader's speech, team discussions, emails, posters, signs, posters, documents, lunch-n-learns, training, seminars, etc.
4. The message must be communicated by at least five different types of people: CEO, the C-Suite, the directors, the managers, the team leaders, colleagues, and coworkers, etc.
5. The message must be delivered on at least five different occasions: at a kickoff session, annually, monthly, quarterly, weekly daily, etc.

The five times rule as calculated is 5 times × 5 learning styles × 5 methods × 5 types of people × 5 different occasions = at least 3125 times! And still, that may not even do it for some. But those who do not get it with at least 3125 instances are not likely to ever get it anyway. They are disengaged. The key understanding here is communicate and over communicate!

Other than the message, the design, and the plan, what's also most important to communicate? The information regarding people, process, and performance. Later in this book, there's a chapter dedicated to each of these three fundamental success factors, plus a "how to" chapter for information on accomplishing an implementation. Performance measurement permits everyone to know what's the status, what's changing, and what's required to take the next step.

At this juncture, I'd like to interject some reality:

- There will be problems.
- There will be conflict.
- There will be issues.
- The design may change.
- The plan will likely need some adjustments along the way.
- All the people may not go along.
- Some of the people will be lost along the way.
- There might occur "the unexpected."
- Be confident, be brave, be resilient, be resolute.

Two other hard spots that frequently appear are related to ideas and mistakes.

- *Ideas*: I often ask the question, "How many ideas does it take to get that one-in-a-million idea?" The real answer I've found is either only one or more than a million. People constantly seek million dollar ideas and most often don't find them. Ideas are to be processed and assessed for innovation. Ideas are inputs to creativity. Concepts realized from ideas are inputs to innovation. If you want innovation, gather millions of ideas. To do this, take in each and every idea before evaluating it. People will be apt to offer up ideas if they know their ideas won't be judged and criticized.
- *Mistakes*: Another question, "Have you ever made an honest mistake?" Also, "Have you ever heard people learn best from their mistakes?" Mistake should be most often treated as learning opportunities, and not as shame-and-blame occasions. Honest mistakes and an atmosphere of experimentation is a salient part of the affordability culture. Leverage mistakes for learning and encourage experimentation where experiments may go wrong (research how many times Edison learned how not to make a light bulb).

Preparedness for change and transformation, according to Blanchard, is key. When assessing and designing the solution, pay keen attention to the following Blanchard suggestions to know:

1. People want information.
2. W.I.I.F.M. (each and every individual wants to know—what's in it for me)
3. Implementation action questions (What do I need to do and when? What help will I get? How long will it take? How will our structure and systems change? Is what we are experiencing typical?).
4. Impact concerns (Is this making a difference? Is it worth it? Are we getting anywhere?).
5. Collaboration concerns (ability to share the good news after progress is made).
6. Refinement and continual improvement: the leaders' role is to facilitate this process modifying from lessons learned ... harvesting ideas to make things even better.

Designing and planning strategies—Blanchard (diagnosis, flexibility, partnering for performance):

1. Expand involvement and influence (outcome: buy-in)
2. Explore possibilities (outcome: options)

3. Select and align the leadership team ... advocates (outcome: one voice)
4. Explain the business case for change (outcome: compelling case for change)
5. Envision the future (outcome: inspiring vision)
6. Experiment and ensure alignment (outcome: collaborative effort and infrastructure)
7. Enable and encourage (outcome: new skills and commitment)
8. Execute and endorse (outcome: accountability and early success)
9. Embed and extend (outcome: reach and sustainable results)

The process:

- Know: overall purpose, strategy, and message for change
- Assess
  - Research and understanding
  - Resources and tools
- Design
  - The design of what it will be
  - The plan of how it will happen
- Implement
  - Successes
  - Understand how to mitigate failures
  - Expect resistance to change
- Maintain
  - Sustain the gains
  - Begin to assess for the "next round"
  - Design when the assess phase is complete

Over the past 30 years, volumes and volumes of information have been published on change and transformation. Affordability embraces many of the proven tools, techniques, and methods. The approaches discussed in this chapter provide only the framework and foundation level for incorporating change and accomplishing transformation. I am expecting to expand this material in the future and elucidate in more depth the entire affordability approach to change and transformation. I'd like to leave this chapter with a couple of case examples.

Case example: SASI technical help desk

Imagine you work for a company that employs hundreds of people. Your group, the technical help desk, comprises 18 people. You come to work, you answer phone calls from customers having problems, and you try to provide solutions. You haven't been trained, you don't have the technical documentation readily available (there are two copies and everyone shares), you don't have equipment to identify the root cause of the technical problem and test your solutions, you don't have engineering support available, and your group is known as the worst place to work in the company. Imagine someone asks you, "How do you know at the end of the day if you've won or you lost?" and you answer, "I know I won if I didn't get yelled at!" How would you go about changing and transforming that environment?

Here's how it happened:

- *Assess*: Time was spent researching and understanding the current state. Resources and tools were identified for use in the operation and as a part of the transformation. A baseline measure of performance was established. Top leadership and group management was involved in caucusing around what is wrong, what can we do, how will we do it?
- *Design*: The top leader, leveraging the crisis condition, creating a dream that stood as a vision and direction, and promising the two managers involved he would support and fund their effort to improve, created an initial design, and developed a plan over a 3-year horizon.
- *Implement*: The leadership team presented the vision, design, plan, and message to the group. The first step was to incorporate some quick wins in the first 30 days of the effort. The victories and celebrations increased morale and motivation and energy was available to make more progress. After the first 6 months, dramatic changes were realized and the plan continued to evolve and adjust according to the conditions and the need.
- *Maintain*: After 2 years of a dedicated effort, the group, now functioning as a team, implemented their own help desk design and now operated with 15 people instead of the original 18. The three other went on to carry the design and plan into engineering, manufacturing, and marketing, thereby starting the transformation of the whole company.

Case example: AT&T/NCR Retail Systems Division

Now imagine you're in a company of around 30,000 people. Your division, the NCR Retail Systems Division, where you serve as one of the 12 executives, provides about 20% of the revenue of your company that was purchased by AT&T for a little more than $9,000,000,000.00 a couple of years ago. You've been going through change and transformation since 1989 when your division opened this new plant where you are now working. The angst of acquisition and merger is among the

Baldrige Examination Summary

Initial Assessment Items	Final % Score	Max Points	Points Rcvd
1.0 Leadership (95 Points)			
1.1 Senior Executive Leadership	40%	45	18.0
1.2 Management for Quality	40%	25	10.0
1.3 Public Responsibility and Corporate Citizenship	70%	25	17.5
Category Total			45.5
2.0 Information and Analysis (75 Points)			
2.1 Scope and Mgmt of Quality and Performance Data/Info	40%	15	6.0
2.2 Competitive Comparisons and Benchmarking	20%	20	4.0
2.3 Analysis and Uses of company-Level Data	40%	40	16.0
Category Total			26.0
3.0 Strategic Quality Planning (60 Points)			
3.1 Strategic Quality and Company Performance Plng Proc	30%	35	10.5
3.2 Quality and Performance Plans	60%	25	15.0
Category Total			25.5
4.0 Human Resource Development and Mgmt (150 Points)			
4.1 Human Resource Management	40%	20	8.0
4.2 Employee Involvement	50%	40	20.0
4.3 Employee Education and Training	0%	40	0.0
4.4 Employee Performance Recognition	30%	25	7.5
4.5 Employee Well-Being and Satisfaction	60%	25	15.0
Category Total			50.5

Baldrige Examination Summary

Initial Assessment Items	Final % Score	Max Points	Points Rcvd
5.0 Management and Process Quality (140 Points)			
5.1 Design and Intro. Of Quality Products and Services	40%	40	16.0
5.2 Process Mgmt - Prod./Svc. Production and Delivery	50%	35	17.5
5.3 Process Mgmt - Business Proc. And Support Services	10%	30	3.0
5.4 Supplier Quality	50%	20	10.0
5.5 Quality Assessment	80%	15	12.0
Category Total			58.5
6.0 quality and Operational Results (180 Points)			
6.1 Product and Service Quality Results	30%	70	21.0
6.2 Company and Operational Results	30%	50	15.0
6.3 Business Process and Support Service Results	10%	25	2.5
6.4 Supplier Quality Results	50%	35	17.5
Category Total			56.0
7.0 Customer Focus and Satisfaction (300 Points)			
7.1 Customer Expectations: Current and Future	30%	35	10.5
7.2 Customer Relationship Management	40%	65	26.0
7.3 Commitment to Customers	70%	15	10.5
7.4 Customer Satisfaction Determination	30%	30	9.0
7.5 Customer Satisfaction Results	50%	85	42.5
7.6 Customer Satisfaction Comparison	45%	70	31.5
Category Total			130.0
Grand Total (1000 Points Maximum)			392.0

NCR: Retail Systems Division 08/24/93

Figure 7.5 NCR Baldrige assessment.

people, a new flagship product is encountering numerous problems when installed in the customer sites, AT&T wants more revenue and profit, and you just saw the results of a Malcolm Baldrige Assessment (see Figure 7.5) that shows your organization is failing (note: the parent company, AT&T, won three Malcolm Baldrige Awards). It is now August of 1993, what do you do?

- *Assess*: Use the results of the assessment to stimulate the emotional energy of the top-line staff for action. Put them to work researching, understanding, and discovering new ways to move forward and make progress.
- *Design*: Design a solution to close 1993 and kickoff 1994 with the next round of improvement initiatives. Update the vision and direction, and put a new plan in place to realize some quick wins and take the organization to the next level.
- *Implement*: Begin the implementation in the last quarter of the year and be sure to include quick wins, big wins, select people, and engage leadership. Follow the eight steps!
- *Maintain*: Fold in the gains and progress and create more change!

Change and transformation is about first and foremost people, and it requires the involvement of the people and the development of people. There is no magic wand, no silver bullet, and no free lunch when it comes to this topic. It is very difficult since it involves paradigm shifts and culture adjustments. But by far, it always includes people. People must be developed to inculcate and infuse change and transformation into a culture.

Executive development

- Executive exposure to cultures that have accomplished change and transformation
- Strategic design and planning of 3–5-year horizons
- Operational design and planning of 1–2 years minimum
- Tactical design and planning of 30–90 days

Guidance team development for those who will be guiding the change and transformation effort:

- *First focus*: Map the value stream, ID waste, future state, plan/design project
- Hierarchy versus dynamic network
- Advise, facilitate, coach, mentor, celebrate progress and accomplishment

Participant development

- Specific training on the design, plan, direction, purpose, reason, and requirements. Including roles and responsibilities and the dynamics of change and transformation.
- Situational training on specific tools and methods to be used.
- Activities, projects, and small quick wins.

Beyond the basics, people have to be prepared, readied, and involved in change and transformation.

chapter eight

Creativity and innovation
From fragments of thought
to prosperity

Creativity

> Creativity requires a confluence of four components:
> Creativity should be highest when 1) an intrinsi-
> cally motivated person with 2) high domain exper-
> tise and 3) high skill in creative thinking 4) works in
> an environment high in support for creativity.
>
> **—Dr. Teresa Amabile**

Innovation

> Innovation distinguishes between a leader and a
> follower. Innovation has nothing to do with how
> many R & D dollars you have. When Apple came up
> with the Mac, IBM was spending at least 100 times
> more on R & D. It's not about money. It's about the
> people you have, how you're led, and how much
> you get it.
>
> **—Steve Jobs**

Creativity is the forerunner of innovation. Creativity is about ideas.
Innovation is about the implementation of those ideas. Organizations
need to seek ideas, a lot of ideas, millions of ideas, if not billions of ideas.
Leadership needs to stimulate the offering up of ideas, gather the ideas,
process the ideas, set the direction for the ideas that will be implemented,
align the proper resources to ensure successful implementation, and
ensure that the ideas will be implemented. Leadership must establish an
environment for ideas and innovation and institute a process for managing ideas from inception to implementation.

There is a line that exists between standardization and innovation,
between static and dynamic. Sometimes, it is the case that innovation
doesn't work well in standardized environments. Of course, if creativity

and innovation are cultural components in the standard environment, then their mutual existence is guaranteed.

Static and dynamic conditions do not easily coexist unless they are designed into a system to coexist and operate as two systems or operate in both contexts as part of the entire system. After being exposed to many design labs, I observed that creative and innovative thought required an environment and atmosphere unlike that of the stable portion of the same organization. However, I have also seen environments where creative and innovative thoughts are integrated into the general workplace. But I do know this, if you want people to come up with creative ideas and innovative ways to solve problems, you must have both (Figure 8.1).

In affordability, it is no coincidence that the foundational layer between change and transformation, and people → process → performance, is creativity and innovation. It is also no coincidence that the next step up from people → process → performance is creativity and innovation. If you want people to change and transform, if you want people to solve the problems of the organization, and if you want people to be engaged in their work, creativity and innovation must exist. One outcome of motivation is creativity, and from creativity comes innovation. Motivate people, you get creativity, maintain motivation, you get innovation.

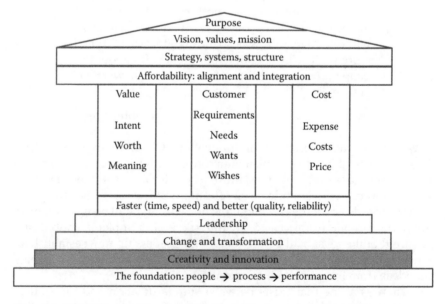

Figure 8.1 The affordability architecture or "the house of affordability."

Ideas

How many ideas does it take to get that "one-in-a-million idea"? I know I've been trying for such a "unicorn," but I have yet to find it. Some people say the answer is one. I say the answer is most probably more than a million. Maybe its somewhere in between. But every organization needs ideas, lots and lots of ideas. Why don't most organizations take advantage of the wealth of knowledge and experience they have and proactively tap into all of their people for ideas? Is it too difficult? Is it too painful? Is it too much of a waste of time?

I've often gone into a room with a group of very smart people. I put a "problem on the table," and I usually get several different types of reactions.

- Silence. People are afraid to offer up ideas.
- An idea or two, and the most important people in the room respond to those ideas.
- A few ideas ... then the others evaluating and commenting on the ideas.
- Several ideas and a debate over whose idea is the best idea.
- Ideas with others "springboarding" off of those ideas to generate more ideas.
- ... and a large variety of other scenarios.

You know what I normally don't get? Someone capturing those ideas, or some way of archiving those ideas for use. Many years ago, I was visiting a Milliken plant in LaGrange, Georgia. In that plant, they operated and maintained an idea system where people input their ideas, the ideas were processed (the people could also check the status of their idea in the system), and if the ideas were implemented, the people or team that submitted the idea would be rewarded (the rewards ranged in size and value according to the entire value of the idea when implemented). I thought that was a great idea, very creative, very innovative, and quite valuable to Milliken. And, you know what? I've never seen anything quite like it anywhere else.

For creativity and innovation, you must get the ideas for creativity, then process those ideas to get to innovation, then innovate for an implementation. But first the environment must be idea friendly in order to get those ideas in the first place.

To get to creativity, people have to do two things: (1) get the ideas and (2) process the ideas. The environment should be set up to allow the free flow of ideas. People should be able to offer up ideas without fear of judgment, evaluation, and critical appraisal. Too often, we are taught to jump to conclusion about ideas and find all the ways the idea will fail (fear of failure). Too often, the ideas are not complete and they are but a fragment of a concept and not the whole concept and are judged to be confusing. Too often, the ideas not being complete, nor being proven, are delayed and never revisited. Too often, the ideas may be similar to others that have failed or not timely or not understood and are criticized (ridiculed). In Dr. John Kotter's book *Buy-In*, he talks about the four strategies for killing great ideas: fear, confusion, delay, ridicule (even character assassination). Our society and culture have readily adopted these responses and we use them in all types of settings (e.g., watch politicians, their favorite responses to kill ideas are fear, confusion, delay, ridicule ... and character assassination). This approach to ideas has got to be stopped. Organizations have to create an environment, a system, and a process for getting ideas, processing ideas, and implementing the ideas of the people, whether it be for problem solving, business development, or organization progress.

Other than the many approaches that have been attempted and used, I have two favorites that have served me well for affordability:

1. Outside-inside, inside-outside, inside-inside: The first time you hear it, it sounds like a line out of a Dr. Seuss book, but it's not. People comfortable in their organization, their establishment paradigm, their "box," may not always like to engage in "out of the box" thinking and discovery to generate new ideas. I like to suggest this to leaders because I serve this purpose all the time, bringing in some of the outside thinking and ideas to the inside for exposure. If done correctly over some time, the inside can be exposed to the outside for additional exposure to similar thoughts and ideas, and eventually the inside that has been exposed can spread the thinking and ideas to the rest of those inside. When at NCR, I sponsored a trip for a busload of associates to Spring Hill, Tennessee, to visit the Saturn plant to enable them to understand their manufacturing approach and their employee training system. I took a vanload of associates to Jacksonville, Florida, to AT&T's Universal Card (A Malcolm Baldrige winner) to help them understand how they measured performance and used it to motivate both leaders and employees. I also took a vanload of associates to Xerox for the understanding of documentation and process. If desired ideas and concepts can come from the outside to the inside for new ideas, and the inside can go out and get some new ideas from the outside, and those "insiders" (exposed from the outside-in and the inside-out) can develop the rest of the inside ... forget the box!

2. In the box? Out of the box? Try a new box?: At times, leaders encourage their people to think out of the box. The "box" is their paradigm of thinking often shaped by the leadership and the organization. To belong and be part of the organization, one should be compliant with the organization paradigm, but being noncompliant and think out of the box is often an approach that is rejected. Thinking out of the box may not work. Thinking in the box is safe, secure, and predictable. Thinking out of the box is risky, uncertain, and can be career-hazardous. So why think in the box or out of the box, create a new box, a new paradigm? If leaders would permit new boxes, people would have a box, and think out of the box at the same time. I've tested this concept on a variety of organizations and I've had success in its principle and results. Create a new box, and support it. New boxes can exist, like Bill Gross's Idealab is one type of example of a new box or Thomas Edison's Menlo Park.

Some of the thinking that created these approaches came from an Albert Einstein quote (Figure 8.2).

Now, once you generate the ideas, what are you going to do with them? How are you going to capture and retain them? If you are able to get them and keep them, how are you going to process them? Many people think that the path to success for creative and innovative ideas are the ideas themselves. Bill Gross, CEO and creator of Idealab, thought the same thing. He thought that the most important thing is the idea. Secondly, he thought the team and execution was a close second. Rounding out his "top 5" were the business model, funding, and timing. After he spent some time researching successful and unsuccessful start-up ideas, both those

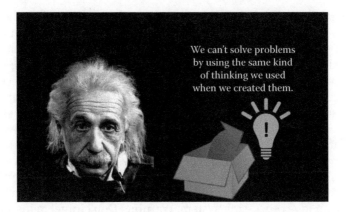

Figure 8.2 Outside-inside, inside-outside, inside-inside thinking.

of Idealab's and those of other origins, he discovered something rather surprising. His results are as follows:

Top five reasons for idea start-up success	Rank
Idea	#3
Team and execution	#2
Business model	#4
Funding	#5
Timing	#1

For success, it goes without saying you need the idea. However, the timing of the implementation is most critical, and the team and execution is second. It makes sense. The idea will be accepted when those needing what the idea provides actually need it. People are the common denominator of the team and execution. The idea, which is mandatory, can only happen when its time has come and when it is implemented by a team that can execute the implementation. If the outcome of an idea is something the customer needs, the business model will adapt to deliver it to the customer. Lastly, funding can accelerate availability, abundancy, and accessibility, but great demand generates funding through revenue. This tells me all ideas can be great ideas: first of all, get the ideas (without judgment, criticism, or evaluation). Capture and keep those ideas because some of them have immediate demand, some of them have emerging demand, and for some, their time may not yet have come.

The idea processing system

Whether it be manual, automated, or a combination of both, an idea processing system permits an organization to record, process, and utilize ideas. It can be designed and implemented as a process. The steps and stages Milliken used (if I remember correctly): (1) get the idea; (2) categorize the idea; (3) management evaluates, rates, and catalogues into a "log" according to the type of idea it is (e.g., some need further investigation, some are not timely and should be stored for later use, some are too expensive, some need technology not yet invented, some are ready to move forward, etc.); (4) the log is processed according to the rules of the log; (5) those ideas to be implemented are put into development and their progression is recorded; and (6) ideas implemented are recognized, the people responsible are rewarded, and the organization reaps the benefits. Invent your idea processing system. Warning: Don't forget to institute it in the culture through change and transformation! (see the previous chapter on change and transformation).

The creative process

The creative process is about an individual idea to an invention. There have been many processes documented that take a fragment of thought to a process, product, or service. The creative process I use within affordability is called the 5 Is of invention (see Figure 8.3). It follows the patterns and stages of many of the inventors, creators, and designers I've known and studied. It often starts with a fragment of thought or an incomplete idea:

- I_1—*Initiation*: The emergence of the fragment of thought or incomplete idea initiates the start of the process. Often, the immaturity level of the idea does not permit definition, nor explanation, nor articulation. Typically, only the originator knows enough to do anything with it. But, to get that "one-in-a-million" idea, each fragment, no matter how small or incomplete, is worth pursuing. Even if it is only used to springboard off of to create other ideas or potential notions to pursue.
- I_2—*Immersion*: The next step requires that the individual or team immerses themselves in everything and anything related to the idea. Discovery requires knowledge, intelligence, information, data, and expertise. Diving into an idea requires that the individuals involved consider as many aspects as possible about the idea. Methods like Lean 3P and TRIZ may be beneficial for starting at this point. Goal: expire the possibilities.

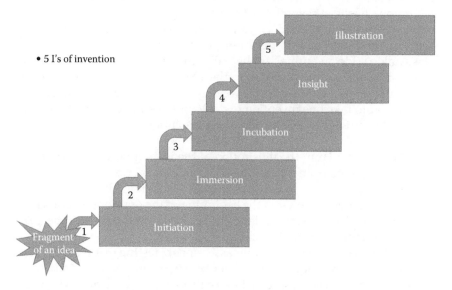

- 5 I's of invention

Figure 8.3 The creative process. The 5 I's.

- I_3—*Incubation*: Immediate results, instantaneous solutions, and all immediate possibilities are not likely. Often, it takes time to incubate an idea and give it time to simmer. The problem with this step in the process is that the duration is unpredictable. I have had ideas that incubate a short time (hours or days). I have had other ideas, like the theory of affordability and this book, that incubated for many years before some understanding and vision emerged.
- I_4—*Insight*: There comes a time, at the end of incubation, that comprehension and awareness comes together and the idea takes on a more malleable form. Quite often, it emerges slowly and clearly, and sometimes there is that "a-ha" moment (warning: those a-has are rare). Once insight occurs, the idea is in complete enough form to put it down in a form that it can be understood and appreciated by others.
- I_5—*Illustration*: If you look back at the great inventors, some of them were artists, some of them were not, but all of them were able to sketch and draw their idea as an invention. The invention occurs when an idea gets to a point of illustration for interpretation and comprehension.

Get an idea, invent something, and use the creative process.

The innovation process

The innovation process is about taking an invention and deploying the solution it provides. There's a saying, "There are many ways to skin a cat." There are many inventions to solve a problem. This process is about taking one of those potential solutions and implementing it as the solution to the problem at hand. At times, two inventions can solve one problem (e.g., problem: how to get from point A to point B. In the 1700s, it was walk or ride an animal. In the 1800s, the inventions of the trains and automobiles solved the problem. In the 1900s, airplanes, jets, and rockets solved the problem. In the 2000s, we may have a teleporter. This process is designed to take one invention to deployment and delivery of the solution. I refer to it as the 5 Ds of deploying a solution:

- D_1—*Design*: From the illustration of the invention, a detailed design can be created; the solution, how the solution is created, how the solution is delivered, and how the solution is maintained. In some organizations, this is known as the beginning of the development process.
- D_2—*Develop*: This phase takes the design(s) and develops the solution (process, product, or service).
- D_3—*Determine/declare*: After assessment, testing, validation, verification, and finalization, a determination can be made and a declaration of readiness can be confirmed and affirmed. Until the solution is ready, the status of readiness should be monitored,

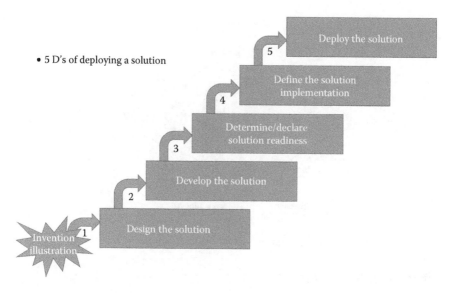

Figure 8.4 The innovation process. The 5 D's.

reviewed, updated, and communicated until it is declared ready for implementation.

- D_4—*Define*: Defining implementation often goes through stages; alpha, beta, limited/select deployment, full availability.
- D_5—*Deploy*: Is the last step in the innovation process. It starts when the solution is being fully deployed through the time deployment ends.

Take an invention, deploy a solution, and use the innovation process (Figure 8.4).

Idea to implementation, creativity to implementation

Here's an idea: how to integrate the 5I creative process with the 5D innovation process (see Figure 8.5). When you start with a fragment of idea and desire to deploy an innovation, you must first be creative, then take the select invention and be innovative. It is often the case that great ideas produce several possibilities for innovation. The point of integration between creativity and innovation provides several possibilities for selecting the solution to produce and deploy. Going from ideas and possibilities to concepts and models is the spot where a sophisticated environment is required. This creative and innovative environment is described and detailed next under innovative capabilities.

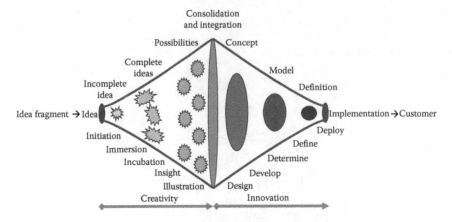

Figure 8.5 Idea to implementation, creativity to innovation.

Innovation capabilities

I was on an airplane, during one of the many flights to and from projects around the world, and I was watching a TED video of Linda Hill on "How to manage for collective creativity." I captured five themes from the video in my notebook (note: I added some of my own comments):

- Unlearn. Think differently than in the way you know now!
- The "a-ha" moment is a myth. It hardly ever happens with an "a-ha"!
- The process is messy. Remember, you're just seeking a "slice of genius"!
- Unleash the talents and passions of many people. That's how Edison really did it!
- Innovative work is exhilarating and scary. Motivation and the unknown in one!

She also had several points of advice to create the creative and innovative environment.

- Leadership is the "secret sauce": creative leadership (not visionary leadership).
- Creating the space where people are willing and able to do the hard work of innovative problem solving.
- Building a sense of community and building the three capabilities.
- Creating an environment where people want to belong.
- Build a "public square" for people interaction.
- All voices are heard.
- Bestow credit in a meaningful way.
- Social architect creating an environment for people.

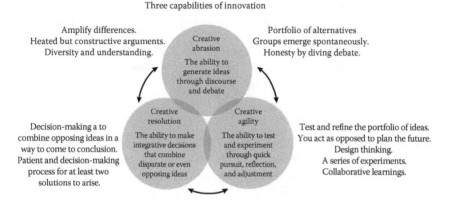

Three capabilities of innovation

Three themes
Collaborative problem solving, discovery driven learning, integrated decision-making

Figure 8.6 Innovation capabilities. (From Linda Hill, Greg Brandeau, Emily Truelove, and Kent Lineback. HBR.org.)

- People closest to the customers for innovation.
- Invert "the pyramid."
- Unleash the potential of the many.
- Leaders: set the stage … not perform on it.
- Create the space of collective genius.

She was also one of a group of people at Harvard Business School who created the three capabilities of innovation (see Figure 8.6). The three capabilities (creative abrasion, creative agility, creative resolution) describe how the behaviors and approaches to thinking, understanding, choice, relationships, and interaction lead to collaborative problem solving, discovery driven learning, and integrated decision making.

The creative and innovative environment

The physical design and operational existence of many creative and innovative environments vary. I've been to such environments at NCR, AT&T Bell Labs, Hoechst, Milliken, Synectics, USC, Georgia Tech, University of Texas Arlington, and others. I've found that they share some specific characteristics:

- A vision, mission, direction, and alignment for creativity and innovation.
- Creative and innovative leaders.
- Creative and innovative people.

- Creative and innovative processes and resources.
- A creative and innovative past, present, and predictable future.

Other than that, they all have their own unique paradigms and practices. But the most important factor, for me, is that leadership encourages, motivates, and supports creativity and innovation.

Creative leadership

The Center for Creative Leadership released a comparative list contrasting the traditional leader and the creative leader. Of course, not all leaders today follow the traditional design; however, not many leaders today follow the creative design either.

Traditional leader	Creative leader
Symbol of authority	Symbol of inspiration
More sticks	More carrots
Hierarchical	Networked
Linear path	Nonlinear path
Plan and execute: launching with 1.0	Iterate and do: living in beta
Sustaining order	Taking risks
Yes or no (clear choice)	Maybe (comfort with ambiguity)
Literal in tone	Metaphorical in tone
Concerned with being right	Concerned with being real
Think like a general or conductor	Think like an artist or designer
Delegates actions	Hands-on driven
One way	Interactive
Close the ranks	Permeable
Follows the manual	Improves when appropriate

And, in order to reap the benefits from creativity and innovation, an organization must have leadership that embraces creativity and innovation, and establishes an environment where creativity and innovation thrives. I have a few unique case examples where I witnessed creativity and innovation from customer input and people input. The first one happened in an environment that embraced creativity and innovation. The second happened outside the area designed for creativity and innovation. The last case example occurred in a war zone.

Case example: NCR 7890 scanner

While an employee at NCR in the Retail Systems Division plant in Duluth, Georgia, 25 years ago, I witnessed and was part of a creative and

Figure 8.7 NCR 7890 scanner.

innovative happening that invented and implemented a product that was driven by customer requirements. NCR had, and still has today, a great deal of expertise in using lasers to read retail bar codes. A loyal customer, The Limited Company, wanted a special bar code scanner for one of its women's garment stores, Victoria's Secret. They wanted a lightweight scanner that could be easily picked up to scan a bar code without pulling a trigger (note: repetitive finger movement causes carpel tunnel syndrome) that would wake up when a bar code was in front of it and then, for energy-saving purposes, go back to sleep when not in use (see Figure 8.7). From that customer input, the 7890 scanner was invented and deployed, and still exists today as the NCR 7893.

Case example: Rockwell Hellfire Missile

While working on a Rockwell project in 1996–1997, I witnessed a fascinating creative innovation that occurred on the factory floor, instead of in the engineering lab where most of the innovation was expected to occur. The Hellfire Missile was being produced in the Georgia assembly plant at that time. After production, it was shipped to Alabama for insertion of the ordinance for delivery to the U.S. military. Once in a while, from the time a pair of missiles were handled in the Georgia plant, shipped to the Alabama plant, and then handled again in the Alabama plant, the clear nosecone on the front of the missile (see Figure 8.8) would get scratched and create a problem for the optics in the missile. The engineers went to work on a solution. They invented a pair of suction devices that could

Figure 8.8 Hellfire Missile.

be fixed to the nosecone and protect it (the cost of the project was nearly $1,000,000.00). The packing and shipping team went to the local Home Depot, bought the materials and tools they needed to construct their design (cost ~$250.00), and created a protective cover that prohibited anything from getting near the clear nose cone. Which solution do you think was better? Both worked. Which solution do you think they used? Yes, the manufacturing solution.

Case example: Cab cart

While I was in Afghanistan in June of 2010 supporting the MRAP program in an effort to increase the speed and quality of the processes of battle damage and repair (BDAR) and upgrades, we were faced with a challenge that required a solution. One of the vehicles being serviced, the MRAP-ATV or M-ATV, was designed as a capsule on a chassis that could be removed for service and repair. Each time the vehicle required capsule removal, a crane had to be brought in, the cab had to be removed and placed on the floor, and then, after the vehicle was repaired, the capsule was lifted and reattached. The problem arose when the steel capsule was off the vehicle it often got in the way and had to be moved, or it had to be moved to be repaired, or both. Each time the capsule had to be moved, a crane had to be brought in. The idea occurred to the service

Figure 8.9 MRAP: the capsule cart.

personnel that if we could elevate and move the capsule around without the crane we could accelerate the process by eliminating several crane moves. A master welder was challenged to invent and build a "capsule cart" that would be mobile and able to support the weight of the capsule. He used his knowledge, experience, creativity, and innovative capability and invented that capsule cart (see Figure 8.9). By eliminating several crane moves, he dramatically increased the speed of BDAR and upgrades.

Creativity is about the idea, and innovation is about the implementation of the idea. Establish an environment for creativity and innovation.

1. Assess: to know
   a. Create an idea creativity environment.
   b. Get more than a million ideas.
   c. Springboard on those ideas to create more ideas.
   d. Store those ideas for the right time.
   e. Illustrate those ideas when the insight occurs.
   f. Process the ideas for implementation for the right time—timing.
2. Design: to strategize and develop
   a. Design and plan.
   b. Create an innovation implementation environment.
   c. Do not be afraid of outside-inside, inside-outside, inside-inside.
   d. Be comfortable with feeling uncomfortable.
   e. Develop solutions.
   f. Release those solutions at the right time.

3. Implement: execute
   a. Know creative leadership and strategize for creativity and innovation.
   b. Learn how to implement and execute.
   c. Turn the people lose on problems and opportunities ... motivate.
4. Maintain: sustain
   a. Keep the creative and innovative environment progressing.
   b. Reinvest in people.
   c. Be aware of elsewhere: See it! Feel it! Do it! Celebrate it!

Creativity and innovation is a competitive advantage!

chapter nine

People

The human factor

> People are born with intrinsic motivation, self-esteem, dignity, curiosity to learn, joy in learning.
>
> **—W. Edwards Deming**

We, the people ... So far, in my own personal "walk of work," I've encountered three careers throughout the past 40 years. The first one was focused in academia with my involvement in mathematics and computer science. The second, in corporate America, beginning as a software development analyst and ending as an executive coach. And the third, my current occupation, as a consultant, concentrating on the areas of leadership and strategy, and also performance improvement. As I came to the end of each of the first two eras, my personal motivation and focus had shifted, and I realized that it was time for a change and even a transformation. However, just like Dr. Deming cited, motivation, self-esteem, dignity, curiosity to learn and joy in learning played a critical role in what I was doing, what I was pursuing, and what I wanted to do next. Over time, I've discovered that many of my friends, colleagues, and collaborators share these attributes, and when their work life experienced a shift in motivation and enthusiasm, they sought out new occupations and professions that fulfilled these aspects in their lives (Figure 9.1).

When I work with organizations around the world, and I observe the people within these organizations, I've seen mixed results when it comes to motivation, self-esteem, dignity, curiosity to learn, and joy in learning as it relates to employee participation, commitment, involvement, and engagement. As I dug deeper and researched the employer–employee relationship, I discovered the results of an interesting study (Table 9.1).

According to the research, most employees are either not engaged or actively disengaged at work. And, in fact, of those who are engaged, only a portion can be considered highly engaged or sometimes termed "high flyers." In addition, as the demographic composition of the workforce changes, their motivations and expectations evolve too. It is imperative to understand what is most valued by the workers. Is it compensation, or prestige, or perhaps autonomy at work? In many cases, the organization will have to adapt their incentives, benefit policies, and retention

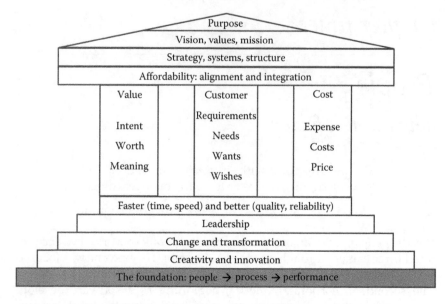

Figure 9.1 The affordability architecture or "the house of affordability."

Table 9.1 Employee engagement

Engaged employees by region	Engaged	Not Engaged	Actively Disengaged
United States and Canada	29%	54%	18%
Australia and New Zealand	24%	60%	16%
Latin America	21%	60%	19%
Commonwealth of Independent States and nearby countries	18%	62%	21%
Western Europe	14%	66%	20%
Southeast Asia	12%	73%	14%
Central and Eastern Europe	11%	63%	26%
Middle East and North Africa	10%	55%	35%
South Asia	10%	61%	29%
Sub-Saharan Africa	10%	57%	33%
East Asia	6%	68%	26%

Source: Gallup World Poll and Gallup Daily tracking survey, 2013 (percentages are rounded).

Note: Results were collected among 73,752 respondents 18 and older in 141 counties and 151,335 U.S. respondents.

strategies for workers who are not just driven by financial compensation. It is not enough simply to recruit able staff. Companies have to make sure that their people are committed, productive, and do not leave after a short period, incurring substantial turnover costs and wasting all previous training invested in them.

Engagement and retention: What follows are several employer considerations compiled from numerous sources over the years that articulate what organizations should do to establish an environment of engagement and retention from the very beginning of the employee's relationship with the employer.

Purpose: Before the employer–employee relationship is determined, an organization should anticipate the needs of future employees by providing a precise definition of the work to be done and an accurate job description. The employer should clearly clarify the employee's role, the company's expectations, and the required skills and qualifications, and the success metrics and measures. The employee's value and purpose should be clear and concise and the performance requirements should be known upfront before the employee begins working day 1. Employees who don't understand the roles and responsibilities become disengaged. Disengagement, especially at the beginning of the working relationship, almost never turns out well for the organization or the employee.

Feedback: The most effective way to provide feedback is through frequent employee–supervisor interaction. Employees need to know that they're performing in alignment with, or better than, the employer's expectations, or contrariwise, less than their performance expectations. Feedback from their supervisors is crucial to correcting deficiencies early in the employment relationship. In addition, feedback satisfies the employee's need for motivation and guidance that reinforces good performance and corrects less than acceptable performance.

Equality: Employees must have the perception of equality in hiring, promotion, training, and retention or they are likely to lose faith in the company. They ultimately become dissatisfied or unmotivated and productivity and performance suffer, as well as morale.

Pecuniary: Employees have their own compensation requirement that allows them to earn wages in exchange for performing their work. Employers must routinely assess their compensation and benefits practices to ensure they are paying their employees competitive wages and providing financial rewards for exemplary service when warranted. Many workers have an intrinsic desire to feel they are making worthwhile contributions to their employer's business. However, satisfying that intrinsic desire isn't enough to sustain an employee's lifestyle. An organization must meet the working population's acceptable wage levels for effective and efficient work production. They also must provide a compensation structure that conveys how much they value their talents and contributions to the organization.

Stability: Employees need to feel that their jobs are safe and stable. Gaining employees' confidence where job stability is concerned means developing a clear communication path, maintaining an open-door policy, and addressing employee issues as soon as they are identified.

What's In It For Me (W.I.I.F.M.): Understanding W.I.I.F.M. for each individual is critical for employers to maintain motivation and momentum. Each employee must understand what's in it for them and how they fit within the organization. Knowing what's in it for them helps to position required changes in a way that is acceptable.

Belonging: One way that leaders can ensure employees are engaged is to ensure that they are thinking, "I belong here." An employee who feels as though she is part of her workplace's community and has an emotional connection to her organization will be more engaged at work. Ways to ensure that everyone on the team feels a sense of belonging include not holding meetings from which certain staffers are routinely excluded and regularly soliciting ideas from everyone on the team, not just the select few "stars."

Enjoyment: For many employees, engagement means having a little fun at the office once in a while. Creating a positive work environment that includes fun ways for employees to interact will go a long way in engaging employees. Whether this means having an occasional birthday or holiday celebration in the office, or a more formal annual retreat, employees will have something to look forward to other than the (let's face it, occasionally monotonous) daily grind.

Alignment: It's important for leaders to acknowledge that their team's efforts are supporting a greater mission or purpose at work. Let employees know specifically how their work is aligned with the organization's mission.

Recognition: Another key component of engagement is employee recognition. Most employees want to be recognized by their managers for their hard work. Leaders who fail to implement reward systems do their employees and their companies a disservice. Whether it's a sizable annual reward or some smaller form of recognition, such as a simple note of thanks, managers should not underestimate the importance of acknowledging employees' successes.

Advancement: Offering plenty of opportunities for employees to build their skills should be a priority for leaders looking to increase engagement. After all, talented employees don't want to stagnate professionally—they want to develop their skills, advance, and thrive. Companies can offer advancement opportunities in a variety of ways, from promotions to in-house training, to college-level courses that the company funds.

With engagement and motivation being very closely related, the following question is often posed: "What is it that keeps people engaged and motivated?"

According to Dr. Rosabeth Kanter, people are primarily motivated and engaged by mastery, membership, and meaning (money is a distant fourth). In other words, people being able to reach a level of mastery in the context of the work they do. People have a want and need to belong and have work be part of their natural daily life and existence. Finally, meaning being related to the significance, importance, and worth of the work being done. She has three recommendations for organizations:

- *Mastery*: Help people develop deep skills. Stretch goals show faith that people can shape the future rather than being victimized by it, and find pride in constant learning. Even in the most seemingly routine areas, when people are given difficult problems to tackle, with appropriate tools and support, they can do things faster, smarter, and better.
- *Membership*: Create community by honoring individuality. Community solidarity comes from allowing the whole person to surface, which means going beyond superficial conformity to know what else people care about. Encourage employees to bring outside interests to work. Give them frequent opportunities to meet people across the organization to help them get to know one another more deeply.
- *Meaning*: Repeat and reinforce a larger purpose. Emphasize the positive impact of the work they do. Clarity about how your products or services can improve the world provides guideposts for employees' priorities and decisions. As part of the daily conversation, mission and purpose can make even mundane tasks a means to a larger end.

Dr. Teresa Amabile, a colleague of Dr. Kanter's at Harvard Business School, and Dr. Steven Kramer recently published the book *The Progress Principle: Using Small Wins to Ignite Joy, Engagement, and Creativity at Work* (HBR Press, 2011). The basis for this book came from a research study I was involved in that was led by Drs. Amabile and Kramer from April 1995 to January 2000 (aka: The Harvard Business School T.E.A.M. Study or The Events And Motivation Study). And, through rigorous analysis of nearly 12,000 diary entries provided by 238 employees in seven companies, they

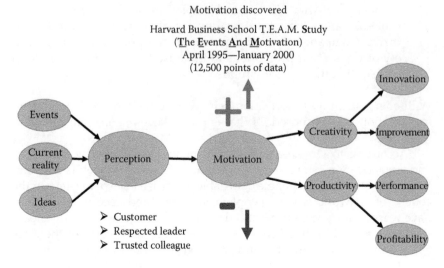

For more information, see *The Progress Principle*, by Dr. Teresa Amabile and Dr. Steven Kramer

Figure 9.2 Motivation drivers and outcomes.

explain how managers can foster progress and enhance inner work life every day and remove obstacles to progress, including meaningless tasks and toxic relationships. It also explains how to activate two forces that enable progress: (1) catalysts—events that directly facilitate project work, such as clear goals and autonomy; and (2) nourishers—interpersonal events that uplift workers, including encouragement and demonstrations of respect and collegiality. An interesting model emerged from my exposure to that study that I continue to use today to explain the relationship of motivation with innovation, improvement, performance, and profitability, as well as with motivation and daily events, current reality, and ideas (Figure 9.2).

The primary outcomes of motivation are creativity and productivity. Creativity drives innovation and improvement of both the people and the process. Productivity drives performance and profitability. Imagine a speech where a CEO gives credit to his/her organization for innovation, improvement, performance, and profitability through creativity and productivity that resulted from motivation (and engagement). The power of motivation is often mentioned, but not always fully understood. I have witnessed great organization with fabulous talents underperform as a result of lack of motivation. On the other hand, I've also witnessed second-rate organizations outperform better organizations sheerly due to motivated people within.

What fascinates me even more is that perception is the primary driver of motivation. It is perception, or belief within one's self, that drives motivation. When we studied the team and people during the Harvard T.E.A.M. study, there were three factors that shaped perception: the person's current reality, events that happened during work, and ideas that were discussed and thought.

Ideally, affordability aspires an environment where all people are engaged and motivated. In the book *The Progress Principle*, Dr. Amabile and Dr. Kramer illustrate three key influences on inner work life: (1) the progress principle (events signifying progress including small wins, breakthroughs, forward movement, goal completion); (2) the catalyst factor (events supporting the work, including setting clear goals, allowing autonomy, providing resources, providing sufficient time, helping with the work, learning from problems and successes, allowing ideas to flow); and (3) the nourishment factor (events supporting the person, including respect, encouragement, emotional support, affiliation). When engagement and motivation are in place, affordability can easily be achieved. Originally, the theory of affordability arose from observing and experiencing organizations where the people were engaged and motivated (the antithesis is also true; affordability did not exist when engagement and motivation was not present). From such conditions and circumstances, the 5P cyclical system and function of success was defined (Figure 9.3).

I have come to realize, however obvious and simplistic, that all work begins and ends with people. Of course, processes are involved, performance is realized, profitability is achieved, and purpose is at the center serving as the reason, the value, the worth, the intent. As we dig into the purpose, we find respect, integrity, and trust (as well as many other

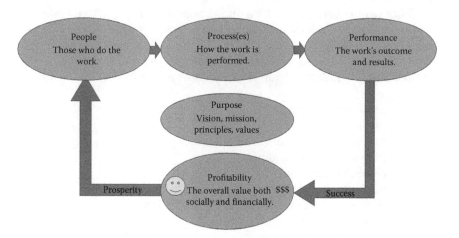

Figure 9.3 The 5P cyclical system and function of success.

values). Principles such as dedication to customer, innovation, and teamwork are but a few. In the purpose also exists the vision, the mission, goals, and objectives (sometimes using S.M.A.R.T. rules: specific, measurable, attainable, relevant, and time-bound) to set direction and align the resources. Additional purpose dimensions can be a part of the fuel for the so-called perpetual engine; communication, design, plan, and schedule for execution.

The people doing the work drive the process and are the primary input that makes this motor run. Engagement and motivation are at the top of the list, so are problem solving, development, and growth. Freedom and autonomy are provided where possible in order to execute experimentation and mitigate mistake. Ideas, concepts, and thinking are critical for creativity and innovation. Rounding off the important list of people input expectations, consistency, stability, standardization, sustainment, future opportunity, and a willingness to let people make mistakes for learning and development. People is the input to process.

Process (explained in more detail in the next chapter) is how the work is performed. This includes resources, tools, machines, information, procedures, methods, means, metrics, and measures. The value-add processes produce either products and services that the customers purchase or they provide support for those value-add processes. Other than products and/or services, the processes output results and outcomes in the form of performance measures.

Performance measures indicate process success or failure and feed the factor of profitability. Profitability is both a monetary and a social facet. This leads to a level of prosperity, which folds back into the people (also including stockholders, stakeholders, and others with involvement in the organization).

With people being the "alpha and omega" of this perpetual system, it is most crucial to understand the requirements and necessities that keep them engaged and motivated. I've collected five lists that state what people want from an employer that affects engagement and motivation.

Ten Things Employees Want Most, by Issie Lapowsky, INC

1. Employees want purpose.
2. Employees want goals.
3. Employees want responsibilities.
4. Employees want autonomy.
5. Employees want flexibility.
6. Employees want attention.
7. Employees want opportunities for innovation.
8. Employees want open mindedness.
9. Employees want transparency.
10. Employees want compensation.

The Intangible Things Employees Want from Employers, by Anne Bahr Thompson, HBR

1. Trust—Don't let me down.
2. Enrichment—Enhance daily life.
3. Responsibility—Behave fairly.
4. Community—Connect me.
5. Contribution—Make me bigger than I am.

Five Factors Every Employee Wants From Work, by Susan M. Heathfield

1. Respect is the fundamental right.
2. Employees want to feel as if they are members of the in-crowd.
3. Employees want to have an impact on decisions that are made about their jobs.
4. Employees want to have the opportunity to grow and develop.
5. Employees do want leadership.

Eight Benefits Employees Wish Employers Would Offer, by Alison Green, Un New: Money

1. Being able to use vacation time without guilt.
2. Being able to use sick time without having to prove you're sick.
3. Being allowed to telecommute when the work allows it.
4. Professional development and training.
5. Flexible schedules.
6. Meaningful roles with real responsibility.
7. Open appreciation.
8. Good management.

CEB's Quarterly Global Labor Market research indicates the top five things employees look for when seeking a new job are

1. Stability
2. Compensation
3. Respect
4. Health benefits
5. Work–life balance

And here is a list for providing for high-potential employees specifically for leaders.

Leaders to provide for high-potential employees (Forbes)

1. Feel valued and respected
2. Sponsor advancement

3. Genuinely invest in growth and development
4. Exposure to people of influence
5. Don't be threatened
6. Encourage risk taking and exploration

Employees want to be compensated, supported, and acknowledged for

- Creating value (through the work that they do, in conjunction with the organization's purpose)
- Problem solving, creativity, innovation (improvement of the organization and their own work)
- Their value, growth, and development (this is their worth and the reinvestment in them)
- Recognition and reward (in keeping with actual performance)
- A share of the success (ongoing profitability and prosperity)

One way to start the ball rolling when it comes to affordability is to put the people to work solving problems. I once heard Dr. Deming say, "People are the solution, not the problem. The process is where the problem occurs. People should be put to work solving the problems of the process." From my own experience, a lot of talent, knowledge, and capability is wasted by companies that don't leverage the advantage of having the people solve the problems of the organizations. A common quote I've heard, "We have people that solve those problems, get back to work and do your job." Instead of, "We do need to solve that problem, what are some possible solutions?" The biggest reinvestment in people with any cost is having the people solve the problems. What follows is the affordability problem-solving process (Figure 9.4).

Although this process has four stages, there is no beginning nor ending. It's a continuous loop:

1. *Assess*: This phase is about research and gaining a complete understanding of the process problem(s) and the root cause(s) of those problems. Resources, tools, fact, data, information, and anything that applies to the process, as well as knowing the current condition and defining the overarching goal. In Six Sigma, it's: define, measure, and analyze. In Lean A3, it's the background, current state, goal, and analysis. The assess phase is a critical input to design and plan.
2. *Design*: This phase is about designing and planning for solution implementation. It includes detailed characterization of the problem and the characterization of the various possible solutions to the problem. It includes picking one solution to pursue and the planning

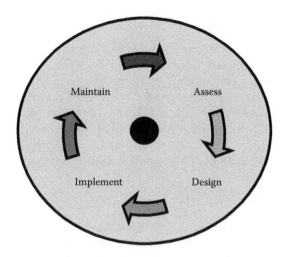

Figure 9.4 The affordability problem solving process.

of how that solution will be implemented. For Six Sigma, it's going from analyze to improve. For Lean A3, it's going from recommendations to plan. Most often, the first two steps of this process are the most critical.

3. *Implement*: This phase is about the execution of the plan and the implementation of the solution. For Six Sigma, the improve step is strongly stated, but it's the implement step that is not as clear. For Lean A3, although the implementation is the institution of countermeasures to mitigate the problem, it can be lost between countermeasures and plan. Emphasis of implementation is a necessity and a must have.

4. *Maintain*: The maintain phase includes maintaining and sustaining the solution, but it also includes continuously checking for improvement opportunities. At this point, problems are seen as opportunities and treasures for discovery and exploitation. This phase starts the cycle all over again.

When people solve problems, it provides motivation and momentum. Forward progress, accomplishment, and achievement create winning teams. Whether you call them short-term wins, quick wins, or small victories, the result is more than the success. It allows people to make the claim "We did it!" Accomplishment fuels accomplishment, just like success fuels success. This approach also requires thinking. And, affordability thinking is an integration of various proven thinking archetypes that work.

Affordability thinking

Affordability incorporates several forms of thinking. The more notable thinking comes from five specific types: GE Six Sigma; Toyota Production Lean; The Fifth Discipline—Peter Senge; The Progress Principle—Dr. Teresa Amabile and Dr. Steven Kramer; and The Three Ms—Dr. Rosabeth Moss Kanter. Affordability thinking leverages these five forms of thinking, integrates affordability's strategic, operational, and tactical aspects, and provides a dynamic, comprehensive approach for thinking about how to deliver value to customers at an affordable cost.

Six Sigma thinking (GE)

1.	Customer:	Delighting customers
2.	Process:	Outside-in thinking
3.	People:	Leadership commitment

Lean thinking

1.	Value:	What the customers are willing to buy and can afford.
2.	Value stream:	The method by which the organization can deliver products and services to the customers.
3.	Flow:	The flow of materials and functions along the value stream.
4.	Pull:	The use and utilization of customer demand for consumption and replenishment method and means creating flow in the value stream.
5.	Perfection:	Elimination of barriers, roadblocks, and impediments that inhibit the value stream. Especially overproduction, inventory, transportation, motion, waiting, overprocessing, defects, and the waste of people ("skill abuse").

Fifth Discipline thinking: Peter Senge's five disciplines

1.	Personal mastery:	People—individuals (every one)
2.	Mental models:	Images—seeing it (including reflection, clarification, "pictures")
3.	Shared vision:	Direction, alignment—future (future state, principles, practices)
4.	Team learning:	Grow and develop—group (collective, teamwork, collaboration)
5.	Systems thinking:	Process → system—enterprise ("How to change systems more effectively, and to act more in tune with the larger processes of the economic and natural world")

Progress Principle thinking—Dr. Teresa Amabile and Dr. Steven Kramer

1. Inner work life:	Progress (positive) and setback (negatives)
2. Progress:	Small wins, breakthroughs, forward movement, goal completion
3. Catalysts:	Setting clear goals, allowing autonomy, providing resources, enough time, assistance/support, learning from problems and successes (wins and losses), allowing ideas to flow
4. Nourishment	Respect, encouragement, emotional support, affiliation

The Three Ms motivational thinking—Dr. Rosabeth Moss Kanter

1. Mastery:	Skills and capabilities
2. Membership:	Belonging and team
3. Meaning:	Purpose and value

Affordability thinking occurs at two levels. One level is at the strategic or organizational level. The second level is operational and tactical level or people and team level.

Affordability thinking (strategic—organizational)

1. Purpose:	Purpose, vision, values, mission
2. Strategy:	Strategy, systems, structure, alignment, integration
3. Affordability:	Value, customer, cost, faster, better
4. Leadership:	Leadership, change and transformation, creativity and innovation
5. People:	People, process, performance

Affordability thinking (operational and tactical—people and team)

1. Purpose:	Value and worth from the customer and organization dimensions
2. People:	Work from the individual and team dimensions
3. Process:	Methods and means from the system and function dimensions
4. Performance:	Outcome and achievement from the qualitative and quantitative dimensions
5. Profitability:	Prosperity from social/cultural and financial/economic dimensions

Affordability's organization purpose

Organizations of affordability have two people dimensions (people and organization) and two perspectives (internal and external). Each of the four specific areas has short-term and long-term goals. People from both internal and external perspectives form a learning community for capabilities, improvement, and the enterprise. The organization internally has the overarching goal of affordability systems (capable processes and continuous improvement) and externally an overarching goal of a value-added partner (profitability, growth, and contribution to society). Affordability is much more than products and services, but dedicated to serving the greater good of the community (Figure 9.5).

Affordability commitment

In order to serve that purpose, commitment should be made by the leadership in four distinct areas: philosophy, people and partners, process, and problem solving. Each commitment represents the various facets of affordability (Figure 9.6).

I've often heard CEOs say, "People are our most valuable asset." Sometimes, I realize they're speaking from a pure cost and expense position. Sometimes, I realize they're speaking from a perception position to build up a viewpoint and mindset. Sometimes, they say this because

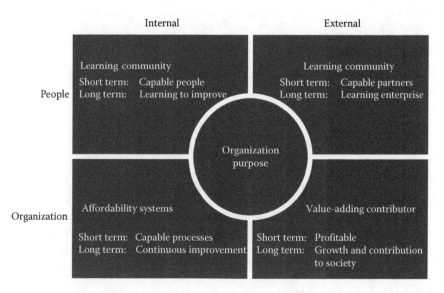

Figure 9.5 Affordability organization purpose. (*Note:* This basic model is adapted from the Toyota way.)

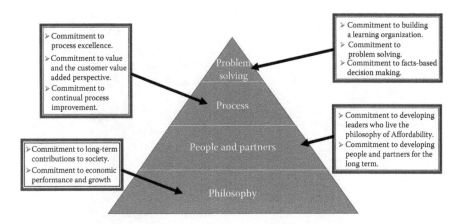

Figure 9.6 The affordability commitment. (*Note:* This basic model is adapted in part from the Toyota way.)

they think it's a good thing to say. But affordability leaders know this and embrace this reality. Yes, it's the people who are truly the most valuable asset. People and affordability fit together quite nicely. A good model for people–affordability can be mapped to Maslow's *human hierarchy*.

Maslow and affordability

If you consider affordability thinking strategically and compare it with two Maslow hierarchy versions, you quickly see that this thinking is a human-based model (Figure 9.7).

- *Purpose*: prosperity and profitability for the basic needs or physiological
- *Strategy*: vision, mission, direction, alignment for security stability and safety
- *Affordability*: a platform for work family supporting families love and belonging
- *Leadership*: establishes, builds, and develops personal esteem
- *People*: the human factor and condition of self-actualization

If you consider Maslow's hierarchy as compared to Ertell's employee hierarchy of needs, again, there's a synonymy (Figure 9.8).

- *Purpose*: Basic tools. The basic tools of affordability.
- *Strategy*: Trust and respect. Values and principles … including integrity.
- *Affordability*: Accountability and authority. For results, outcomes, accomplishment.

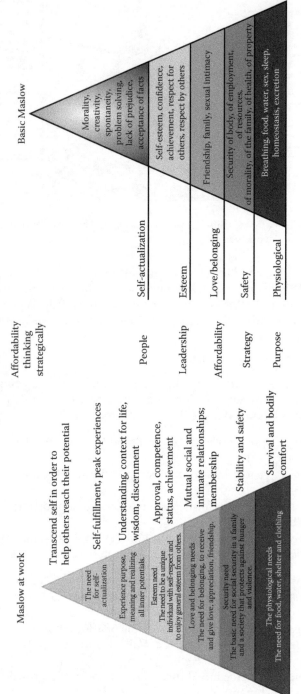

Figure 9.7 Maslow at work and Basic Maslow as aligned with affordability thinking.

People
Maslow, affordability, customer satisfaction

Employee satisfaction leads to customer satisfaction
(and big profits) January 19, 2011, by KEVIN ERTELL

Figure 9.8 Maslow's hierarchy of needs as compared to Ertell's employee hierarchy of needs.

- *Leadership*: Confidence. Organization certainty, self-reliance, conviction.
- *People*: Growth. And the development of skills, capabilities, and competencies.

Finally, with Maslow's hierarchy and employee engagement thinking, you begin to see how engagement, motivation, and performance play a role.

So in essence, affordability is about people. Engaging and motivating people to solve problems and perform. Growing and developing people for ability, competence, and capability. Creating teams of people able to overcome and best the competition. Establishing successful, prosperous, and profitable communities of people for the betterment of mankind. This is the ideal future state, it is not an absolute, but it is a condition of aspiration (Figure 9.9).

The people of affordability come in five specific forms and profiles:

1. Customers
2. Partners
3. Suppliers
4. Teams
5. Stakeholders

Customers

Customers of an affordability organization are the people who purchase the products and/or services the organization delivers. They are the reason for existence, the reason for being, and the focus of the main purpose.

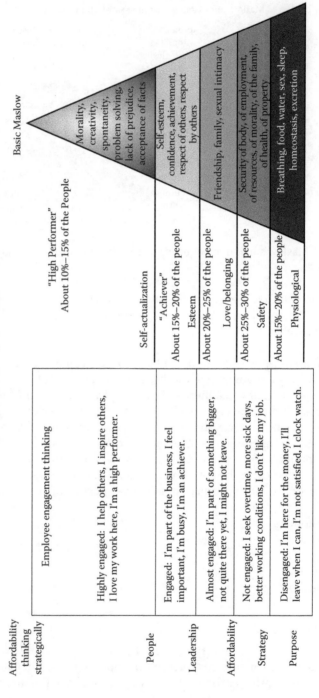

Figure 9.9 Affordability thinking aligned with Maslow's Basic hierarchy.

Partners

A partner is a collaborator. Someone or some group working with you to deliver the expected outcome of the work you're doing. In my business, I often rely on partners for placing me with clients and customers. Often, my customers serve as partners to refer me to other customers. The best partners have common values, principles, goals, and aspirations. The best partners will provide support, ask for and receive support, and compliment the work of the people in the organization. Some partners can also qualify as suppliers.

Suppliers

Suppliers provide knowledge, tools, methods, materials, and other services that assist in delivering the expected outcome of the work you're doing. Suppliers and partners do the things that you do not do for the customer or do the things you cannot do for the customer. In my line of work, I use suppliers for technology tools such as IT systems, computing and calculating methods, and communication devices. I also use suppliers for physical services such as meeting rooms, transportation means, and meal providers. Suppliers provide what is needed for the value stream in addition to what the organization provides or needs in order to provide.

Teams

If you are the only person in an organization able to perform a particular function, and the only one capable of performing that function, either you're a company of one or you're in an organization that has at least one critical failure point. A company of one is understandable, but a single point of failure in organizations of more than one is inexcusable. Organizations of affordability, being customer focused, prepare mitigating responses to potential failures. The best organization construct for failsafe affordability comprises teams and teamwork systems.

As teams grew in popularity beyond the playing field, a lot of literature was published to support this growth. One of my favorite books on teams in the workplace is the *Team Handbook* by Peter Scholtes (now in its third printing). His process excellence approach to management using teams integrates nicely with affordability. His "forming, storming, norming and performing" (I add reforming and change responding) description of the four team phases of team maturity are on target from a practical and pragmatic point of view. He covers roles and responsibilities, teams at work, teamwork, problem solving, improvement, dealing with conflict, and provides additional tools and techniques to support the team effort.

Affordability teams are cross-functional for process and systems problem solving, and multifunction for team roles and core responsibilities using rotational leadership and revolving assignments for basic team tasks. For example, a manufacturing team will sometimes require engineering, quality, finance, human resources, installation service and maintenance, and sales and marketing to help them solve system problems with the products being produced. Also, for example, the nursing staff of a healthcare team may use doctors, physicians, assistants, technicians, housekeeping, food service, and patient advocacy to solve a problem in the care process. The manufacturing team example might use a rotational approach for the safety lead, the quality lead, the communications lead, the training lead, the materials lead, and the team lead of the group. Likewise, the nursing staff could implement the same approach as the manufacturing team with a 90-day rotation plan for each role and position. Dynamic teams have proven robust, flexible, and durable (if done right). If anything, teams fulfill the human need of belonging and membership, and the ability to celebrate success, the "We did it together!" and the learning from setbacks, challenges, and problem solving. Celebration and recognition: it is the human presence that drives affordability.

Even if a process is fully automated, supposedly independent of any human involvement, that process needs some form of human involvement in maintenance and sustainment in order to continue to operate. All processes involve some form of human participation, however great or small, for improvement. Process improvement and performance is controlled and regulated by the people involved with the process. This means that a process, designed with human participation, is under the control of the people involved. The process will produce what the process is designed to produce, but will not meet production performance if the people involved throttle the process in any way. For example, if a machine is stamping metal to produce a product, and the humans involved fail to maintain and service the machine, the machine will ultimately fail due to wear. Another example, if a banding machine does not have its banding material replenished when necessary, the process of banding will fail. Yet another example, if medical instruments are not cleaned properly, their use and utilization will ultimately cause disease. People drive the process resulting in performance, leading to profitability where prosperity is returned to the people.

Stakeholders

Stakeholders, composed of partners, suppliers and teams, have a vested interest in the outcome of the work that people do. Some stakeholders are internal or part of the intrinsic activities of the organization. While other stakeholders are external, or an extrinsic, or an extension of the organization, outside the normal course of activities.

Case example: NCR

When I joined NCR on January 5, 1985, in Columbia, South Carolina, the company had been in business more than 100 years. It had a very success-ful past and still clung to its past success and archaic corporate structure. We were an organization of departments and divisions, often compet-ing from within for resources and permission. Instead of being customer focused, we were internally focused primarily on profit and profit alone. Most often, the method used for reducing cost was to layoff or downsize. In 1987, I was promoted to a corporate strategic planning position in Dayton, Ohio, NCR's headquarters at that time. The "view from the top" was no different than the view from the bottom, out in the field. In 1989, I was asked to move from my current position in H.Q. to a new opportunity in Duluth, Georgia. The Retail Systems Division had designed and planned to open a new plant based on the Just-in-Time (J.I.T.) manufacturing phi-losophy and style of production. What I witnessed there during the early 1990s was truly a transformation that took place over the next 5 years. The team-based approach (that was described earlier) for J.I.T. using demand to provide fulfillment upon customer consumption was revolutionary and new to the company. Especially where the customer was the focus of activ-ity, value, value stream, flow, pull, and perfection was the way we were going to do business. The transmogrification of the teams was a phenom-enal sight to see. Going from a dog-eat-dog competitive and sometimes coercive environment, to an open integrated system and process approach to production was, for lack of a better term, awesome. Our satisfaction improved, our on-time delivery improved, our quality improved, our prof-itability improved, and the people improved. At the end of 1994, from the beginning of 1990, our customer satisfaction went from 72.1% to 85.6%, our associate satisfaction went from 52.0% to 87.0%, our delivery performance went from 60.2% to 93.0%, our overall system quality (with 100% functional and operational quality) went from 76.2% to 99.7%, our production capacity went from 72,000 to 98,000, our development cycle time went from 24–36 to 12–18 months, our revenue went from $220MM to $282MM, and our profit went from −$2MM (loss) to $5MM. During the first week of January, all nonmanagement employees, for the first time in the history of the com-pany, received a performance bonus check for 3.04% of their 1994 salary.

Case example: Store Automated Systems, Inc. (SASI)—technical help desk (refer Figure 9.10)

I was brought in by the leadership team (i.e., the four owners of the com-pany) in December 1997 to begin working with, as they put it, the worst department in the company, the technical help desk. Their job was to field technical service calls and resolve customer issues. On the first day I

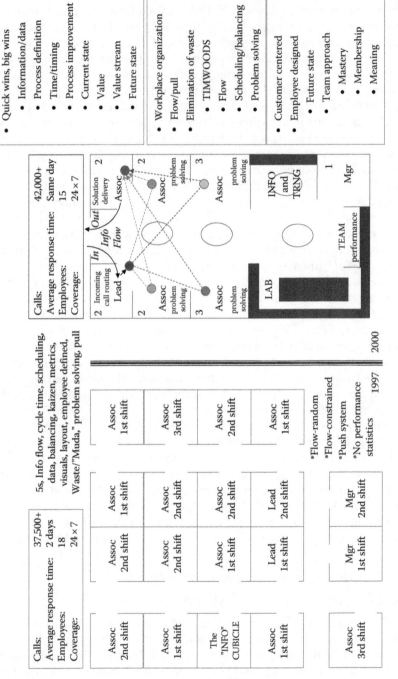

Figure 9.10 Help desk: technical service and support transformation from 1997–2000.

started working with them in January 1998, I was told the only way to get out of this group was to quit or die. To say the least, motivation was low and teamwork was nowhere to be found. In fact, instead of even being a department, it was a group of independent individuals existing in what I call a "cube farm." The integration of group members into a team happened after they could collaborate on solving problems, improve their workplace using new tools and techniques, and improve their environment by designing a better place to work.

Case example: GM stamping plant

Over the years, I always heard how auto workers were difficult to work with. My first exposure came in the early 1990s when I took a busload of manufacturing workers to Spring Hill, Tennessee, to tour and observe the way the Saturn automobiles were manufactured. Quite a great, eye-opening, experience, and very enjoyable. My second exposure happened when I was asked to come to the GM Doraville plant and train UAW workers about quality, Lean, and Six Sigma. GM had decided to in-source a stamping operation and get away from sourcing doors, hoods, panels, and other stamped body parts from different parts of the country. Before the stamping operation got under way, the plant manager wanted to be sure that her workers were trained, knowledgeable, and capable. To make a long story shorter, the stamping plant opened with a two shift team configuration and paid for itself within a year.

Case example: Virginia Blood Services

Virginia Blood Services is the organization that supplies most of the whole blood and blood components to healthcare systems in southern Virginia. Their system collects, processes, and distributes whole blood and blood components. Within 3 days (yes only 3 days), three teams of 8–10 people each performed three rapid improvement events at three points within their system:

- *Team* 1: warehouse—supporting the setup and support of the blood drives
- *Team* 2: collections—collecting the blood at the blood drives
- *Team* 3: lab—processing the blood that comes into the facility from the blood drives

This coordinated effort, across three points of the system, yielded an "annuity" of $34,000.00 in just 3 days. This annual saving was due to the elimination of waste, reduction of defects, and improvement in the processes involved. In fact, it increased their participation rate and increased their "flow" (donor flow that is).

chapter ten

Process

Work work work!

> All work is a process.
> If you can't describe what you are doing as a process, you don't know what you're doing.
>
> **—W. Edwards Deming**

My understanding of process was forged by three perspectives of learning.

- Knowledge:
  - *Describe*: Be able to describe the process.
  - *Know*: Know the process.
  - *Transfer*: Be able to transfer process knowledge and understanding to others.
- Experience:
  - "Don't try to describe the ocean if you've never seen it" (Jimmy Buffett).
  - Don't try and describe the process if you've never seen it.
  - Go to the Gemba (Toyota Production System: Lean Learning).
  - Go out and experience the process first hand.
- Expertise:
  - *Process level*: From the internal, independent viewpoint.
  - *System level*: From the internal and external, interdependent viewpoint.
  - *Enterprise level*: From the enterprise, all-encompassing viewpoint. Everything, up to, and including, all customers and markets where the value is being delivered.

Working for NCR during the late 1980s and the early 1990s, and also for AT&T Bell Labs during the early 1990s, I gained a great deal of knowledge and experience. However, I did not fully get my expertise to an operable level until I encountered customer after customer and project after project over the past 20 years. And I must admit, I gained knowledge, experience, and expertise continually. After serving 85 organizations and being involved in more than 150 projects, I am more than convinced, I'm convicted that Dr. Deming was right. The problems in the process are for the people to fix, so that the performance increases. The foundation layer of affordability must include process (Figure 10.1).

Affordability, other than being about integrating value, customer, and cost for continuous improvement, is also about the relationship and reason for purpose, people, process, performance, and profitability, including success and prosperity. I often refer to this relationship and reason as "the 5Ps, a cyclical perpetual system and function of success" (see Figure 10.2). The way it works, people drive the process, resulting in

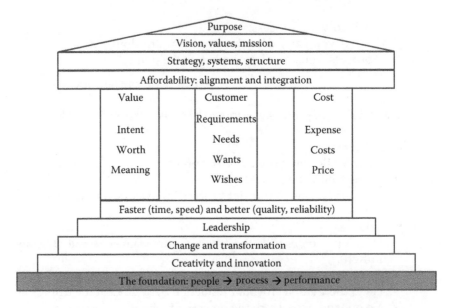

Figure 10.1 The affordability architecture or "the house of affordability."

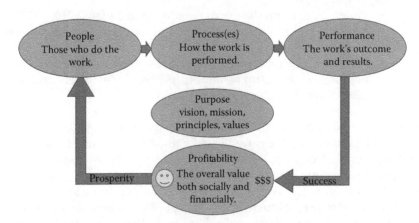

Figure 10.2 The 5Ps: A cyclical perpetual system and function of success.

performance, and the outcome is profitability and prosperity, all starting and ending with people, encompassing purpose.

Purpose: The purpose is made up of the values, principles, vision, mission, and all of the other fundamentals and essentials that guide, direct, and align the organization. Customers and markets, value, costs, and financials can also serve to align people with the purpose.

People: People provide input to and drive the process. No matter how well the process is designed, or how well the process is implemented, people do have dominion over the process. The 5Ps start and end with people for a reason. It is the people who best provide the power and fuel for this 5P "engine" to run.

Process: The process takes people as the primary input and produces outcomes and results that integrate into performance.

Performance: The accomplishments and consequences of performance is the level of success and achievement.

Profitability: Profitability is the end result of performance, both good and bad. Within profitability there are social and economic dimensions. Profitability of people and monetary profitability. Prosperity is an outcome of the combination of people profitability and monetary profitability that provides the input to the people to keep the cycle going perpetually.

At a strategic level, the work that I do is illustrated with the 5Ps. It's basically a process, and in this particular case, it's an ongoing process that lasts as long as the purpose, people, process, performance, and profitability last. There is some importance in documenting processes so as to make it easier to describe them to others, so that they know what you're doing (just like Dr. Deming says). When documenting a process, there are several steps required. What follows is an ordered and summarized seven-step approach:

1. *Observe*: Watch the process while it's operating. Take note of material, machine(s), tools, techniques, methods, and people involved. Make sure that all activities, actions, and movements are noted. Involve people in the process in helping you understand what's taking place. Take pictures or videos if possible to capture the real work.
2. *Document*: Write down the stages and steps and create flow charts, illustrations, and drawings to describe the process. Use pictures and videos for documentation too.
3. *Validate*: Review your process documentation with the work team involved in the process for accuracy and authenticity. Make necessary changes and adjustments.
4. *Deploy/implement*: Put the process "in play" by getting the work team to execute the process.
5. *Utilize/use*: On an ongoing practice, check the process for validity and use the process where necessary.
6. *Sustain/maintain*: Make adjustments or changes when necessary.

7. *Continuously improve*: Continually assess and evaluate the process for improvement needs. If possible, insert several performance points of measurement for assessment and evaluation purposes.

One way to document a process is using the classic supplier-input-process-output-customer (SIPOC) approach (see Figure 10.3). It's a simple method for recording who the suppliers are, what the suppliers provide as input to the process, what the process does, how the process works, what the process outputs, and who the customer of the process happens to be. Here are the descriptions of each of the components:

- *Suppliers*: Suppliers can be external or internal providers. Suppliers can provide material, information, resources, tools, machines, maintenance, or service(s). Their output is the processes input. Suppliers provide everything that is needed to complete the process.
- *Input*: Input is what is required for the process to meet customer requirements, standards, and deliver to the customer what the customer wants, when the customer wants it, at a reasonable price.
- *Process*: A process is composed of activities that are related and organized to take the input and transform it into output in a way that meets customer requirements and standards. The activities in the process must work together in an orderly fashion to meet a common goal, objective, and standard.
- *Output*: Accomplishment and achievement of the process is the output. The output is the value that the customer seeks.
- *Customer*: The one who takes the process input as part of a system or as part of a transaction.

The SIPOC can be even further dissected, studied, and assessed from an anatomical, structural, and functional viewpoint. The SIPOC anatomy diagram provides an in-depth view of all the constituents that make up a quality process. In addition to the core SIPOC items (i.e., suppliers, input, process, output, customer), ownership, management, communication, and performance measurement should be included as mandatory essentials in process documentation. Knowing process ownership

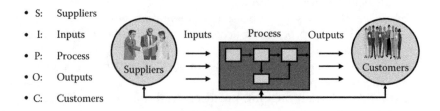

- S: Suppliers
- I: Inputs
- P: Process
- O: Outputs
- C: Customers

Figure 10.3 The basic S.I.P.O.C. elements.

and management provides the people with who's accountable and even responsible for the process. Knowing how the process is performing is important information for process change, improvement, maintenance, and sustainment. Communication throughout the process between suppliers and the process, between the process and the customer, and even within the process keeps all of the people abreast of status, changes, and needs (Figure 10.4).

Another tool for describing the process is the value stream map (VSM). It's often used at a strategic level or sometimes called the 30,000 ft level to illustrate the flow of material and information through the process from the suppliers to the customer(s) over time. In essence, it's a complex SIPOC from suppliers in the upper-left-hand corner providing input to the process, with output flowing to the customer in the upper-right-hand corner of the diagram. It details the steps, the data and information of each step, and the critical time factors (e.g., value-added time, nonvalue-added time, cycle time, changeover time, lead time, Takt time, etc.). Visuals, like the VSM, are powerful tools for describing and articulating how the process works and operates. Such tools can be used for training, reference, and improvement. Although this is a simple example here, more complex and detailed maps are used for identifying and eliminating waste, creating a view of the future state of how the process should function and be included in process improvement team events (Figure 10.5).

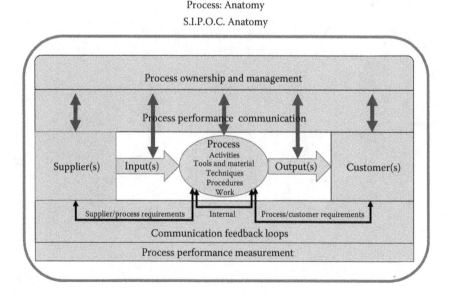

Figure 10.4 Anatomy of a process using the S.I.P.O.C. framework.

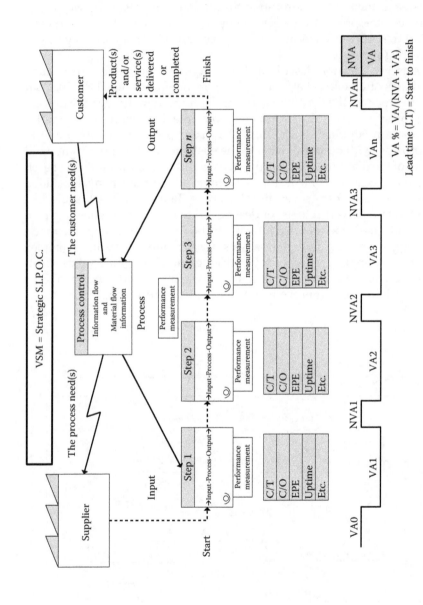

Figure 10.5 Value Stream Map (VSM) framework.

Another tool for describing a process is the flow chart. Using specific functions, for specific types of activities, a process can be mapped and published for various uses (see Figure 10.6). Flow charts are often used with VSMs for detailing steps at an operational or tactical level. They can provide a view of the order of the work being done, the relationship of the work activities, and the work flow of the process. Although flow charts have been available for use for a long time, many organizations do not bother to use them for a variety of excuses. It can be time consuming to document processes using flow charts, and too often, long-time-to-produce is the excuse for not flowcharting a process.

Flow charts can be depicted as swimlane charts or swimlane diagrams. The following example illustrates a process that crosses four departments and takes three phases to complete. The flow chart for this process is arranged in such a way that it clearly shows the work of each department and when it's done. The swimlane chart takes on the image of a swimming pool, where each function stays in its swimlane and performs the process work according to the workflow explained by the flow chart (Figure 10.7).

I've used two different methods to detail a SIPOC and illustrate its "pieces and parts." Process SIPOC illustration method 1 lists the process steps, the suppliers, the inputs, the outputs, and the customers. It also includes the customer and supplier requirements, as well as the customer and supplier performance metrics and measures (Figure 10.8).

Process SIPOC illustration method 2 is similar. However, it uses a more graphic and illustrative approach to detailing the SIPOC. Either method is acceptable, and both serve the purpose of providing SIPOC itemization (Figure 10.9).

At the worker level, the flow process chart (see Figure 10.10) uses symbolism, sequential steps, and time to explain the work being performed by the worker in the process. This is a common industrial engineering approach to study work and describe what the worker does. This tactical approach is not used as much today as it once was. Old methods can often be revitalized to reach new and better levels of performance.

I'd like to walk through a simple example of how all of these tools and methods can be used. We'll use a fictional character Mrs. K. Mrs. K is of great value to her family. She has a husband and two children. Most of her weekdays are very busy. On a typical day, she gets up at 06:00 in the morning, makes breakfast, goes to work, comes home, makes dinner, spends some family time, and retires at 10:00 in the evening in order to get 8 hours sleep. To say the least, she is very organized, stable, standardized, and disciplined during the week (although her weekends are very

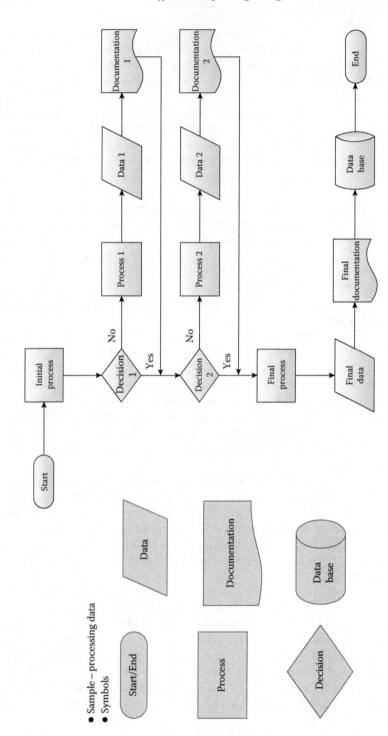

- Sample – processing data
- Symbols

Figure 10.6 Process flowchart.

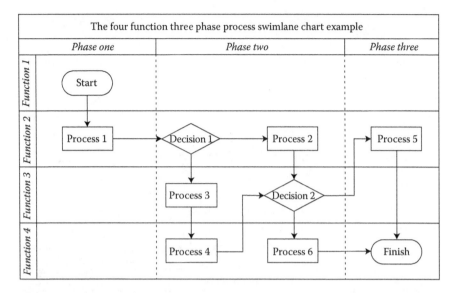

Figure 10.7 Process swimlane chart.

Supplier(s)	Input(s)	Process	Output(s)	Customer(s)
----------	----------	Step 1	----------	----------
----------	----------	Step 2	----------	----------
----------	----------	Step 3	----------	----------
----------	----------	Step 4	----------	----------
----------	----------	⋮	----------	----------
----------	----------	⋮	----------	----------
----------	----------	⋮	----------	----------
----------	----------	⋮	----------	----------
----------	----------	⋮	----------	----------
----------	----------	Step *n*	----------	----------
Requirements	Supplier performance measurement	Process performance measurement	Customer performance assessment	Requirements
----------	----------	----------	----------	----------
----------	----------	----------	----------	----------
----------	----------		----------	----------
----------				----------

Figure 10.8 Process S.I.P.O.C. illustration method 1.

different than her week days). At the "30,000 ft level," her average week-day is illustrated by the following VSM. You might say she's very efficient, and I might add that the way she organizes the household, many of the required functions and chores seem to have been nicely distributed among the other family members (Figure 10.11).

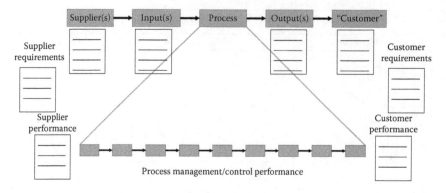

Figure 10.9 Process S.I.P.O.C. illustration method 2.

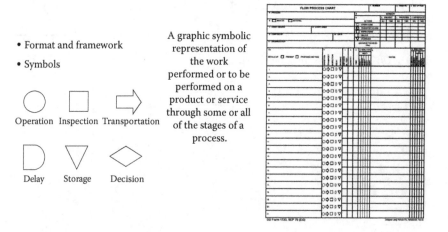

Figure 10.10 Flow process chart.

If we look deeper into her day, say "breakfast," we can see her work-flow (a flow chart is practical here and a swimlane chart would not be applicable since she is the only person performing this work). From the start of the breakfast time until the end of the breakfast time, each and every step is clear and the order of the work steps are distinct (Figure 10.12).

The breakfast flow chart can be further documented using either of the SIPOC illustration methods. I chose method 1 for this example, but method 2 can be populated with the same detail (Figures 10.13 and 10.14).

Within the breakfast process, making coffee is a key step. The flow process chart provides a very detailed view of the activities and actions required for coffee production. You can see, at every level, from Mrs. K's

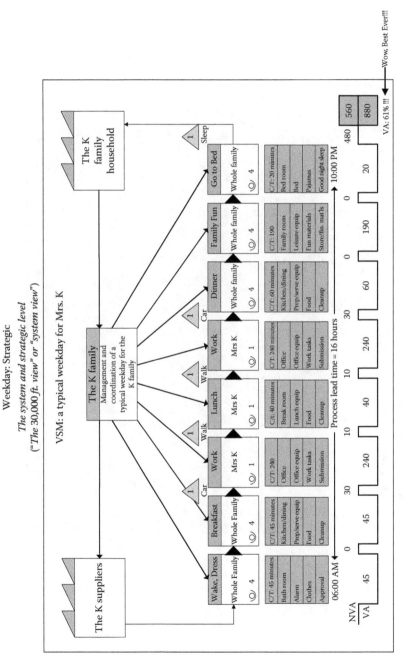

Figure 10.11 Mrs. K's value stream map.

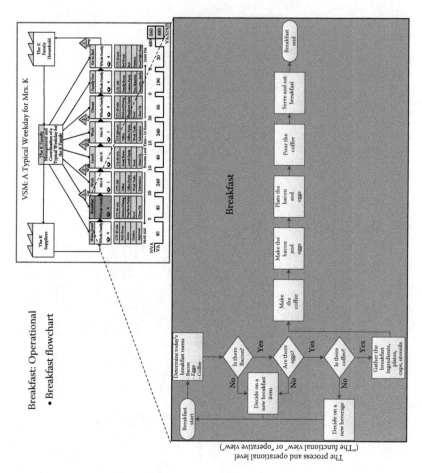

Figure 10.12 Mrs. K's process flow chart.

Supplier(s)	Input(s)	Process	Output(s)	Customer(s)
• Food	• Bacon	1. Determine	• Good breakfast	• Mr. K
• Refrigerator	• Eggs	menu	• Good bacon	• Child – 17 years
• Cooking Equip.	• Stove	2. Verify	• Good eggs	• Child – 18 years
• Stove	• Pans/utensils	materials	• Good coffee	• Mrs. K
• Pans	• Coffee	3. Gather		
• Utensils	• Brewer/pot	materials		
• Coffee Equip.	• Filters	4. Make coffee		
• Brewer	• Water	5. Make food		
• Pot	• Plates	6. Plate food		
• Eating Equip.	• Cups	7. Pour coffee		
• Furniture	• Utensils	8. Serve and eat		
• Utilities	• Cleaning Supplies	9. Cleanup		
Requirements • Good food • Ample food • Tasty meal	Supplier performance measurement • Mrs. K's satisfaction	Process performance measurement • Time • Quality	Customer performance assessment • On time • "yummy"	Requirements • Good food • Warm food • Healthy

Figure 10.13 Mrs. K's process S.I.P.O.C. breakfast (operational): method 1.

day, to breakfast, to making coffee, process mapping and documentation can be completed at every level, strategic level, to the operational level to the tactical level, and from the enterprise view, to the operational view, to the tactical view, right down to the actual work view (Figure 10.15).

The power of pictures and videos

As you can see, process documentation does not imply documents only. Images and illustrations have more impact on people than documents full of text. With today's technology, there is no reason to provide process documentation in text form only. In fact, I have a good personal example that shows a process in action from the before, to the after. I conducted a 5S event in my garage. I took a picture before I started, and a picture after a day's worth of work (see Figures 10.16 and 10.17). One afternoon, I went into my garage, and you can see from the before pictures, I had quite a mess on my hands. That first day, I sorted, set in order, and swept the garage. The next day, using a few shelves I assembled, standardized, and set the place up to be sustained. There were some shocking results that I did not notice until after I viewed the photos. For example, in the before picture in Figure 10.16, did you notice the car that appeared in the after picture? (the car was there all the time). In the after picture in Figure 10.17, do you see the Subaru car top carrier? … it is in the before picture under the black and white checkered blanket in Figure 10.16. My point is that photos and videos have a profound impact for the sake of proof of concept, profound learning, and

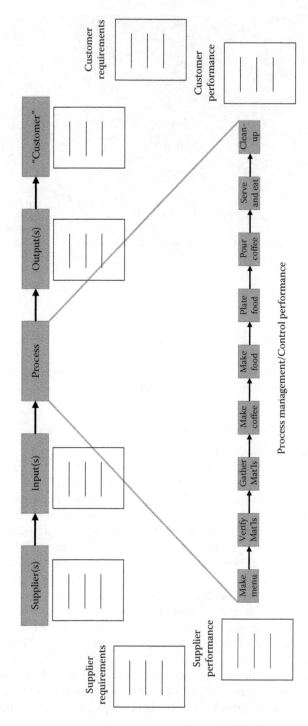

Figure 10.14 Mrs. K's process S.I.P.O.C. breakfast (operational): method 2.

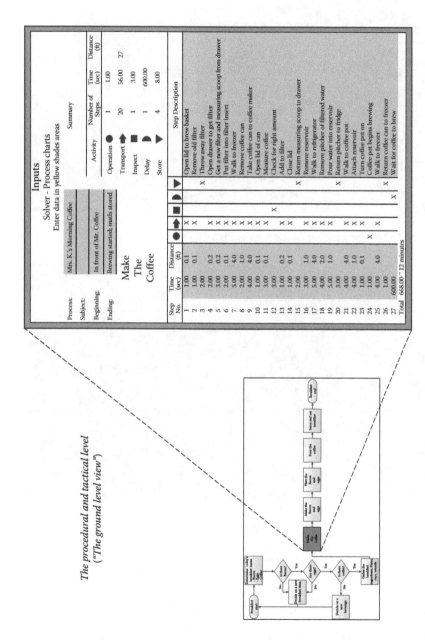

Figure 10.15 Mrs. K's flow process chart–make the coffee: tactical.

Figure 10.16 The left side of the garage. Before and after.

Figure 10.17 The right side of the garage. Before and after.

perception change. Use images, illustration, photos, and videos for visuals as much as possible when creating process documentation!

Additional support methods and models

I have never been a big believer of always having to invent a new wheel or always having to be on the leading edge for new methods and models. There are a multitude of excellent process improvement and quality frameworks have appeared over the past 30 years. Before I started providing process improvement services to other organizations, I read, researched, experienced, and learned about proven methods and models. I was first exposed to J.I.T. production in 1989. I began studying Dr. Deming's principles in 1990 and met him shortly thereafter. My involvement with ISO certification occurred in a 1990 certification award. I was able to use the Baldrige criteria to assess our NCR retail systems leadership team. And, I was exposed to, and learned from, Bell Labs after AT&T purchased NCR in the early 1990s. Although I consider these approaches to be closely related to quality, I position them with process due to their content that goes far beyond quality and addresses all the organization dimensions and dynamics for improving work and performance. The process is the work and the output of each process is performance, and every organization must be well orchestrated for maintaining process excellence and practicing the ever-vigilance and actions required to continuously improve their processes. The ones that best pair with affordability are Dr. W. Edwards Deming's 14 Principles for Management (1986), The Malcom Baldrige Criteria from the National Institute of Standards and Technology (NIST, 1988), The Toyota Philosophy and Universal Principles, and The International Standards Organization (ISO) or The International Organization for Standards, Seven Quality Management Principles (ISO 9001:2015). As you read through each framework, the synonymy and parallel with affordability comes clear. In addition to the affordability principles, the complimentary four are as follows.

Dr. W. Edwards Deming's 14 principles for management

1. Create constancy of purpose toward improvement of product and service, with the aim to become competitive and to stay in business, and to provide jobs.
2. Adopt the new philosophy. We are in a new economic age. Western management must awaken to the challenge, must learn their responsibilities, and take on leadership for change.
3. Cease dependence on inspection to achieve quality. Eliminate the need for inspection on a mass basis by building quality into the product in the first place.
4. End the practice of awarding business on the basis of price tag. Instead, minimize total cost. Move toward a single supplier for any one item on a long-term relationship of loyalty and trust.
5. Improve constantly and forever the system of production and service, to improve quality and productivity and thus constantly decrease costs.
6. Institute training on the job.

7. Institute leadership (see point 12 and Chapter 8). The aim of supervision should be to help people and machines and gadgets to do a better job. Supervision of management is in need of overhaul, as well as supervision of production workers.
8. Drive out fear, so that everyone may work effectively for the company (see Chapter 3).
9. Break down barriers between departments. People in research, design, sales, and production must work as a team to foresee problems of production and in use that may be encountered with the product or service.
10. Eliminate slogans, exhortations, and targets for the work force asking for zero defects and new levels of productivity. Such exhortations only create adversarial relationships as the bulk of the causes of low quality and low productivity belong to the system and thus lie beyond the power of the work force.
    a. Eliminate work standards (quotas) on the factory floor. Substitute leadership.
    b. Eliminate management by objective. Eliminate management by numbers, numerical goals (substitute process goals). Substitute leadership.
11. Remove barriers that rob the hourly worker of his right to pride of workmanship. The responsibility of supervisors must be changed from sheer numbers to quality.
12. Remove barriers that rob people in management and in engineering of their right to pride of workmanship. This means, inter alia, abolishment of the annual or merit rating and of management by objective (see Chapter 3).
13. Institute a vigorous program of education and self-improvement.
14. Put everybody in the company to work to accomplish the transformation. The transformation is everybody's job.

The Malcom Baldrige criteria from the National Institute of Standards and Technology (NIST)

1. *Leadership*: How upper management leads the organization and how the organization leads within the community.
2. *Strategic planning*: How the organization establishes and plans to implement strategic direction.
3. *Customer and market focus*: How the organization builds and maintains strong, lasting relationships with customers.
4. *Measurement, analysis, and knowledge management*: How the organization uses data to support key processes and manage performance.
5. *Human resource focus*: How the organization empowers and involves the workforce.

6. *Process management*: How the organization designs, manages, and improves key processes.
7. *Business/organizational performance results*: How the organization performs in terms of customer satisfaction, finances, human resources, suppliers and partner performance, operations, governance, and social responsibility, and how the organization compares to its competitors.

The Toyota Philosophy and Universal Principles published in The Toyota Way by Dr. Jeffrey K. Liker

1. Base your management decisions on a long-term philosophy, even at the expense of short-term financial goals.
2. Create a continuous process flow to bring problems to the surface.
3. Use "pull" systems to avoid overproduction.
4. Level out the workload (work like the tortoise, not the hare).
5. Build a culture of stopping to fix problems, to get quality right the first time.
6. Standardized tasks and processes are the foundation for continuous improvement and employee empowerment.
7. Use visual controls so that no problems are hidden.
8. Use only reliable, thoroughly tested technology that serves your people and process.
9. Grow leaders who thoroughly understand the work, live the philosophy, and teach it to others.
10. Develop exceptional people and teams who follow your company's philosophy.
11. Respect your extended network of partners and suppliers by challenging them and helping them improve.
12. Go and see for yourself to thoroughly understand the situation.
13. Make decisions slowly by consensus, thoroughly considering all options; implement rapidly.
14. Become a learning organization through the relentless reflection and continuous improvement.

The ISO or the International Organization for Standards

Seven quality management principles

- *QMP 1—Customer focus*: The primary focus of quality management is to meet customer requirements and to strive to exceed customer expectations.
- *QMP 2—Leadership*: Leaders at all levels establish unity of purpose and direction and create conditions in which people are engaged in achieving the organization's quality objectives.

- *QMP 3—Engagement of people*: Competent, empowered, and engaged people at all levels throughout the organization are essential to enhance its capability to create and deliver value.
- *QMP 4—Process approach*: Consistent and predictable results are achieved more effectively and efficiently when activities are understood and managed as interrelated processes that function as a coherent system.
- *QMP 5—Improvement*: Successful organizations have an ongoing focus on improvement.
- *QMP 6—Evidence-based decision-making*: Decisions based on the analysis and evaluation of data and information are more likely to produce the desired results.
- *QMP 7—Relationship management*: For sustained success, an organization manages its relationships with interested parties, such as suppliers.

Case example: Anixter St. Petersburg kitting process

The Anixter kitting process in St. Petersburg started out as, what I called, a trick-or-treat process. The kitting team would make the boxes for an order, lay them out on the floor, then walk by each one, dropping a piece of material called for in the kit down into the box. This continued until all pieces of the kit were dropped in every box. The team would then gather up the boxes, place them in a pallet, and shrink wrap them for shipping. As the team matured and learned more advanced approaches to kitting, they designed a system that was quite creative and innovative. Without any additional expense to the company, they set up a kitting line with seven stations (note: their research told them the kits they were building up had anywhere from 4 to 14 pieces, no more, no less). As a kit order came in, they could configure the kitting line and set up the material at each station to fulfill the order. In addition, they were able to place two photos in of each kitting person that showed them what the kit should look like when they get it at their station and what it should look like when it is passed to the next station. This was a configurable kitting operation that had sight quality validation at every step in the process. After 20 minutes of kitting, they would stop, take a 5-minute break, and rotate stations to avoid complacency and fatigue. Through implementing this design, they reduced their kit defects down to practically zero. When they did happen to find a defect, the team would engage in a problem-solving process event. And it was all defined, implemented, and maintained by the team.

Case example: Siemens environmental control systems, Alpharetta, Georgia

Siemens plants always have process documentation for their workers. However, I did encounter one plant where the documentation went from volumes of text to photos with volumes of support text. Siemens Environmental Control Systems plant in Alpharetta, Georgia, produced a quality product. However, it was their intent to move to a higher level of performance. To make another long story short, they introduced photos to the production line as a part of process documentation and dramatically affected their quality, efficiency, and cost.

Conclusions

Describe your process in easily understood ways and methods. Manage your process using metrics and measures that clearly communicate performance. Solve problems in your process as part of the work of the people. Maintain and sustain your processes using affordability tools and techniques.

chapter eleven

Performance
What is the score? Are we winning or losing?

What gets measured gets done.

> **—Attributed to many gurus, including Peter Drucker, Tom Peters, Dr. W. Edwards Deming, Lord Kelvin, et al.**

Performance, and performance alone, dictates the predator in any food chain.

> **—A Navy Sea, Air and Land Team (SEALs) saying**

Performance can be good or it can be bad. There have been times, after noticing the absence of a performance system, I've asked clients, "How do you know if you're winning or you're losing?" I had one response, more memorable than the others, "Well, we didn't get yelled at today." That response came from a group lead in a service group that I began working with in January 1997. In January of 2000, her response was, "We're winning! I can even tell you from a Customer, Employee, or Supplier perspective, or a Time, Quality, or Cost perspective!" A newly acquired client of mine just answered the same question with, "We just don't know." I've found, often is the case, even where a lot of performance measures are captured and communicated, the people of the organization aren't cognizant of the actual performance of the organization. In numerous instances, I've often been told that it's whether or not we're making money.

Although performance is the last element I cover in this book, it's the first in my mind when I assess an organization. Usually, I'm invited into an organization due to a crisis, failure, malfunction, or critical need of some sort. When I first begin assessing the situation, I try and discover the performance, process, and people factors (in that order). Personally, I know when I'm being successful due to customer involvement, time (time dedicated to delivering value), quality (as a result of the type of work I'm engaged in), and profit (as a result of cost taken out of revenue).

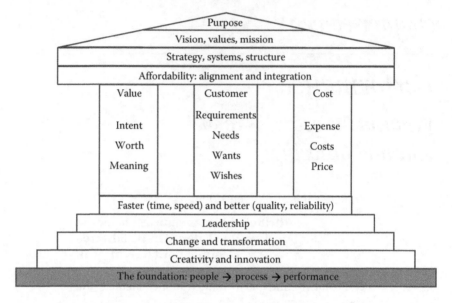

Figure 11.1 The affordability architecture or "the house of affordability."

But, too often in the workplace, people are evaluated on things that don't contribute to value, customer, and cost. Instead, they are too focused on the boss, the boss's boss, and the internal desires and perceptions of the organization. Performance is measured as an output of process and work (see Figure 11.1). Affordability uses qualitative and quantitative measures to evaluate performance as a result of people working in process(es) that provide value. Customer satisfaction, people satisfaction, and supplier/ partner satisfaction are all qualitative measures that can be collected, reviewed, and reported. Time, quality, and cost are all quantitative measures that indicate the performance of the process. Keep in mind too that unless a person is assigned to a process requiring only one individual the outcome of processes are, for the most part, team results achieved from process results. Measure the process, not the people. Put people to work solving process problems. Two points of principle 10 from Dr. Deming's 14 Principles for Management imply that quotas and management by numbers should be eliminated and substituted with process goals and leadership:

- Eliminate work standards (quotas) on the factory floor. Substitute leadership.
- Eliminate management by objective. Eliminate management by numbers, numerical goals (substitute process goals). Substitute leadership.

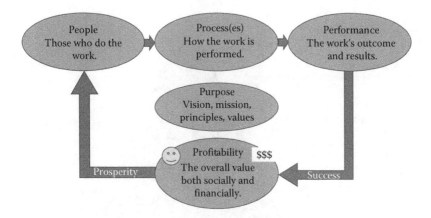

Figure 11.2 The 5Ps, a cyclical system and function of success.

If you want to measure people performance, measure top leadership's results on value, market share growth, customer base growth, profitability, prosperity, and overall all the growth factors of the organization. At the people level, process performance results of customer/people/supplier/partner satisfaction, time, quality, and cost, which are mostly attributed to team accomplishments and achievements.

Considering the 5Ps (see Figure 11.2), key performance indicators can be established at each point:

- *Purpose*: values, vision, mission, goals, strategy, structure, and systems
- *People*: capability and capacity, growth, and accomplishment
- *Process (including system and enterprise perspectives)*: customer satisfaction, people satisfaction, supplier satisfaction, partner satisfaction, time, quality, and cost
- *Performance*: total process performance, customer and market growth, competitive results, and best-in-class status
- *Profitability*: overall value, financial profitability, social profitability, business success, and overall prosperity

The 5P enterprise view level provides an archetype people can align with, follow, and be motivated to engage, support, and contribute their efforts. At the 5P system view, people can understand how everything they do integrates and comes together to provide value and meet customer requirements. At the 5P process level, people can directly apply their efforts to accomplish and achieve success. The 5Ps can be applied at the strategic level ("enterprise"), the operational level ("systems"), and the tactical level ("process," "work") for an architectural perspective of how all the pieces fit for the organization.

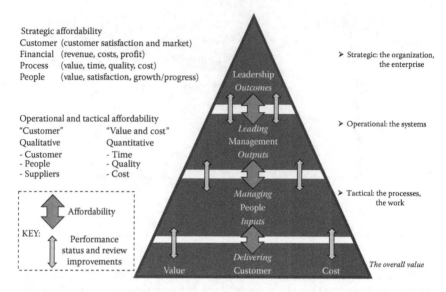

Figure 11.3 Affordability's performance pyramid.

In affordability, the overall value and all performance is founded upon the three pillars of value, customer, and cost, and resultant from strategy and leadership with purpose, faster and better accomplishments, change and transformation, and creativity and innovation complete the architecture of the "house of affordability." The three pillars of afford-ability provide for the integration of value, customer, and cost for continu-ous improvement. Affordability's measurement system is illustrated by affordability's performance pyramid (see Figure 11.3).

Affordability's performance pyramid illustrates how performance can be cascaded and communicated throughout any organization. It's a dynamic design using a bottom-up as well as a top-down approach. Starting at the customer, or bottomline layer, input flow upward to the tactical layer of people performing the work and delivering the products and/or service to the customer. The people and teams of people provide process outputs in performance to the layer of process and system man-agement who in turn service the people by managing the process and systems. The operational layer, or management layer, provides system outcomes to leadership who return performance by leading the organiza-tion. So, leading then managing then delivering flows from leadership, through management, through people to the customer. Performance sta-tus, review, and improvement communication use the catchball technique path to pass information and communication upward and downward in order to maintain linkage. This pathway can also be used for establishing goals and objectives in an effort to maintain policy deployment throughout the organization. (Note: this approach also includes the customer layer performance communication and also for requirements inputs and value responses.) Each level is configured in a way that captures, communicates, and executes according to measures that best fit each level of the organization. Following is a template for performance measurement and reporting, reviewing, and improving:

1. Top level—strategic (organization and enterprise): In concert with, and aligning to Kaplan's and Norton's "balanced scorecard," the metrics and measures for organization performance are categorized into four primary areas and can be measured as follows:
   a. Customers
      i. Satisfaction/delight
      ii. Value delivered
      iii. Conformance to requirements
   b. Financial
      i. Financial
      ii. Strategic
      iii. Operational/tactical
   c. Process
      i. Management (standards, compliance)
      ii. Performance (time, quality, cost/expense)
      iii. Continuous improvement (creativity/innovation)
   d. People
      i. Satisfaction (meaning)
      ii. Competency (mastery)
      iii. Growth (membership)
2. Middle levels—operational and tactical (systems and processes): qualitative and quantitative
   a. Qualitative
      i. Customers
      ii. People
      iii. Suppliers/partners
   b. Quantitative
      i. Time
      ii. Quality
      iii. Cost
3. Bottomline level—customer: the overall value
      i. Customer satisfaction
      ii. Conformance to requirements
      iii. Compliance with standards
      iv. Time: delivered on time
      v. Quality: excellence value
      vi. Cost: price—reasonable, acceptable, competitive

How do you communicate performance to your organization today? Annual reports? Speeches? E-mails? Dashboards? Scorecards? Internet or intranet websites? Posters? Banners? Daily and weekly reports? Word of mouth? Or, not at all, leaving rumor and innuendo to do the talking? The people and teams should have a place to go to see how their team is doing. Whether you call it a scoreboard or something else, the title really doesn't matter. The key attributes need to be accurate, consistent, maintained, true, stable, used, utilized, clear, concise, simple and straightforward, and integrated from top to bottom, bottom to top, linked and connected with the qualitative and the quantitative metrics and measures that display the status and can be addressed for improvement purposes.

At the people, team, tactical, and operational levels, scoreboards can be created for communication throughout the organization. A sample of such a scoreboard is illustrated in Figure 11.4. Using a three-section, or three-panel approach, people, process, and performance can be conveyed. The people panel features the individuals, their competency and capability, their skills and knowledge, and their activities and news. The process panel features the process of course from a value stream and process map perspective, as well as what's changed and improved in the process and what problems have been solved and what problems are being solved. The performance panel contains qualitative and quantitative performance data with comments on performance and potential needs for improvement.

Operational and tactical scoreboards can be used daily, weekly, monthly, quarterly, annually, or whenever the need arises to discuss wins,

Status board reviews (15–20 minutes)

- Status
- Updates/changes/improvements
- Needs

 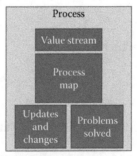 

*At the operational level process is "system."

Figure 11.4 Performance communication boards, for communicating performance of people, team, process, and performance. Used at operational and tactical levels.

loses, improvement, or opportunities. The people and the team should be accountable and responsible for their own scoreboard and use it for leadership and management reviews and even for customer and supplier/ partner visits. With Lean practices, Gemba walks might include the scoreboards. MBWA practices could include the scoreboards with management walking around. I used the scoreboards on "quality tours," where the top leadership of the organization spent 2 hours, 10:00–12:00 visiting each production team for 15 minutes obtain performance status, activities, plans, and needs.

Affordability also institutes scoreboards at the strategic level (see Figure 11.5). By linking aims, goals, targets, and objectives throughout the organization, a consistent view of performance can be conveyed. Using a framework, such as the balanced scorecard (Kaplan and Norton, 1996), a strategic scoreboard, linked to the operation and tactical scoreboards can be employed, produced, and maintained. It demonstrates leadership's integrity of measuring and managing the organization based on the same measures and metrics used at all the other levels. It also establishes a discipline of providing the right success measures that build trust. Leadership embracing a consistent performance measurement system throughout all levels also expresses and communicates respect in that "they are using the same performance metrics that we are measured on" (i.e., "We're in this game together. Playing on the same field.").

As case examples, I'd like to share two distinct instances of performance measurement I've been exposed to, which come from very vastly different settings. The first one, E&M Atlanta, produced and delivered products to their retail customer worldwide. The second one, Gwinnett County Tax Commission, provided a government service of tax collection for the eventual benefit of the citizens of Gwinnett County Georgia. One example had a more formal performance measurement system, while the other had a very simple, straightforward method of gauging performance. Both worked, and both stand as good examples of measuring performance and excellent results and outcomes.

Figure 11.5 Performance communication boards for the strategic level.

Case example: E&M Atlanta 1990–1994

The core design center for the scoreboard examples in this chapter evolved from this example, and my work with Bell Labs, Harvard Business School and Quality Gurus (Dr. W. Edwards Deming and Philip Crosby) supplied the remaining design nucleus for the affordability performance scoreboards. The people → process → performance approach to affordability originated from my work at NCR and AT&T from 1989 to 1995. In Figure 11.6, you can see that the people/process/performance team scoreboards were physically and strategically located in each of the eight team areas. The rolled-up consolidated plant/division-level scoreboards were located in the center of the plant, just inside the associate (our name for employees) point of entrance and exit, in the main hallway. The teams maintained their scoreboards, and the management and leadership team maintained the plant/division scoreboards. Every Friday, from 10:00 until 12:00, an executive quality tour was held for each team to report status, updates, improvement, and needs of the people, process, and performance of their team. Since each of the eight locations had 15 minutes to report out, the executives had to mostly listen. By Monday, the plant/division scoreboards were updated with the new information. All associates could see whether they were winning or losing. In fact, when customers were taken on plant tours, they too could see the performance, as well as the suppliers and partners who visited the plant. Anybody who entered the plant could see, at a glance, how we were doing. At first, the leadership had a lot of fear and anxiety in giving everyone a peek at our traditionally held, private data. But after several months, it became apparent—it was to our advantage to show how well we led, managed, and operated the plant. The coup de grâce came on the first Friday of January 1995 when every associate in the plant walked in and received their first-time-ever performance bonus check payout for the results that they achieved in 1994.

Case example: Gwinnett County Tax Commission

In 1984, Katherine Meyer was elected to the office of Gwinnett County Tax Commissioner. She ran on a simple platform that she maintained for the 26 years she served and was continually re-elected to that position: customer focus, faster, better, and less costly government. At the time she was originally elected, the tax commission office was very internally focused and not at all customer friendly. The tax commission office provides to county residents the value and value streams of billing and collection of automobile and property taxes for use in county government (fire, police, parks, utilities, recreation, etc.). At the end of 1984, the population was 229,270, and when she retired in 2010, it was 808,304 (it had grown by

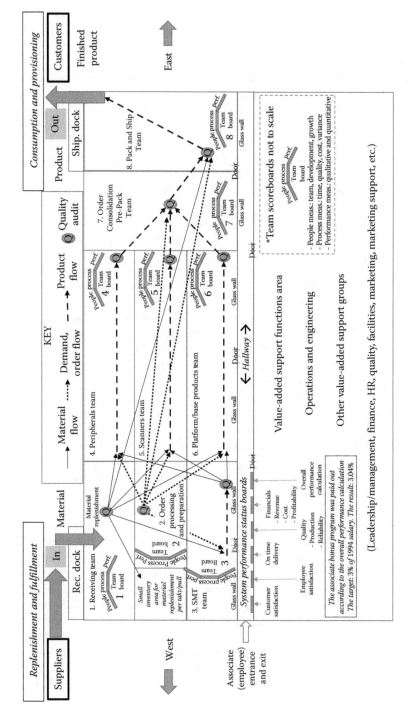

Figure 11.6 NCR, retail systems division, Duluth, Georgia, 1989–1995. Floor layout for the value-added and value-added function.

more than 350%). In 1984, it was not uncommon for people to wait several hours, if not all day, to get a license plate or license plate renewal. In 1984, tax bills were fraught with misinformation and that organization, supposedly serving those citizens, was plagued by service process defects. The baseline performance metrics and measures of that organization could most probably not get much lower.

When I moved to Gwinnett County in 1989, I volunteered to serve on her business advisory council and continued to do so until she retired. Her performance philosophy was based on customer focus, employee growth and development, faster, better, and less-costly services through process improvement. She was re-elected six times before she retired. And, after being elected back to office four times, the population in the county was 596,296 and the customer focus was apparent by the way the employees treated the citizens in the tax offices (they went from one central office to five offices logistically located throughout the county with very little cost for implementation). The waiting times in the license plate offices were averaging only 8–12 minutes (I know personally because every year I chose to stand in line instead of purchasing my plate renewal either over the phone, via the mail, or, soon to be released at that time, by way of the Internet), and the tax bills and notifications contained very few errors.

From 2000 until the time of her retirement, the population of the county had grown by more than 212,000 people. In that time, 2000–2010, she did not have to hire additional personnel, she did not have to increase her budget, and she was able to put into the county coffer around 35% more in collections. Throughout the years, her model of performance for affordability in government was right in front of me in action: customer satisfaction, employee satisfaction, time, quality, and cost. As the department evolved from 1984 until 2010, the rest of the state took notice and quite often used Katherine for input and recruited her employees to gain knowledge, competency, and performance. Well done Katherine!

Performance is a universal reality in all of our lives. Even with young children, to borrow a long-lost phrase, know "the thrill of victory and the agony of defeat." When I was coaching youngsters, many years ago, I used a common idiom before each contest: "When we win, we celebrate! When we lose, we learn." After each game, I would ask the same question: "What are we going to do?" (When we won, we celebrated. When we lost, I asked them, what did we learn? They knew the power of performance.) Let no one fool you in this "P.C." world we live in. Know this; *the score is important.* Everyone in the organization needs to know if they're winning or losing. In this day and age of sophistication, complication, and technology, we often overwhelm ourselves by measuring too much, and not keeping focused on customer, people, suppliers, partners, time, quality, and cost. We need to have more affordability victories or learn how to achieve affordability.

chapter twelve

How to

Being busy does not always mean real work. The
object of all work is production, or accomplishment, and to either of these ends there must be fore-
thought, system, planning, intelligence, and honest
purpose, as well as perspiration. Seeming to do is
not doing.

—Thomas A. Edison

How to achieve affordability is akin to "eating the elephant." We all know
that eating the elephant is achieved by taking one bite at a time. However,
I rarely see a design and plan for achieving such a feat. Affordability is
achieved by aligning and aiming at the integration of value, customer,
and cost. This chapter provides one approach to accomplishing that mean.
This approach is the compilation of 25 years of exposure to numerous
similar endeavors. After being involved with more than 80 organizations
and way more than 150 projects, I've discovered some key actions and
attributes that lead to affordability.

As you've gathered from all the previous chapters, affordability is
accomplished by understanding and knowing the customers of the target market, providing value by way of meeting their needs, wants, and
requirements at a price that meets their expectations and means. The
understanding of the "what-it-is" should be clear by now. It's the "getting
there" that looms as the biggest challenge.

Over the years, several popular approaches have been utilized to
achieve the alignment of affordability. Motorola and GE used the Six Sigma
approach. Toyota (TPS—Toyota Production System) uses their approach
that has come to be known as Lean. Lean and Six Sigma have been combined to create Lean Six Sigma. Kaplan and Norton prescribed a balanced
scorecard, while Michael Hammer re-engineered the corporation. Over
time, we've encountered PDCA, PDSA, QC, TQM, QMS, BPM, TOC, LSS,
and many other tools and techniques that serve as toolsets and toolkit for
achieving affordability. At this juncture, I must tell you, it's not the tools
and techniques that get you there, it's the you and the me, and the we that

accomplishes the change and transformation required. I've observed that implementations of affordability share at least 14 characteristics:

1. Research and understanding of the current state and the opportunities for improvement.
2. A purpose. A need. A desired state. Possibly realized via crisis, or "impossible dream."
3. A customer-centered perspective and focus with an operational reality and realization.
4. A direction that serves the organization as a rally and focal point, with a sound message.
5. Great ideas articulated in terms of a how to for implementation.
6. A leader who decides to pursue these ideas defines an "initiative," and creates change.
7. An engaged leader who understands the science and dynamics of leading transformation.
8. A leader with a strong leadership team with influence and organization effectiveness.
9. A collaborative purpose, vision, direction, and alignment from leadership.
10. An organization instituting motivation, creativity, innovation, growth, and communication.
11. A strategy and plan with clear goals and objectives designed for execution.
12. Process centric systems thinking with tools and methods for success.
13. Recognition of achievement and accomplishment … "success celebration."
14. Perhaps most important of all, people motivated, driven, and inspired to "make better."

As we know, transformation is a journey and pursuit, not a project, nor a scheme, nor a "strategy du jour." It requires purpose, leadership, strategy, design, planning, tools, methods, resources, and most of all, the people.

The first full stage of implementing affordability may occur in months; however, it usually requires 3–5 years, and oftentimes longer for an organization to make the transition. After in-depth assessing and designing, early wins, short-term wins, or as I call them, "quick wins," should occur in the very first month of implementation and throughout the first year. Realization of strategic long-term gains and positive outcomes plus attainment of new paradigms and standardization of processes should begin to appear in the first couple of years. Many large organizations often need 5–8 years to realize their initially targeted gains.

The implementation of affordability requires four phases: assess, design, implement, and maintain. The affordability phase line (see Figure 12.1) details those four phases and briefly describes what the leaders, people, program, resources, and tools typically do and when they typically do it. The first half of the affordability phase line is composed of the seven steps required for the assess and design phases. The second half of the affordability phase line is primarily the implement phase and the beginning of the maintain phase. The affordability curve is flat throughout the assess and design phases and accelerates throughout the implement and maintain phases. The stars and the curves depict where events, learning, growth, and progress take place, as well as how variation from enthusiasm to complacency is likely to occur. The path from the current state to the future state increases and completes the first round of the institution of affordability.

Needs, musts, wants ...

Affordability can only take place if there's a need, must, and want to drive the energy for attainment of a better functioning organization. The most successful implementations I've seen start with a "dream" or a "crisis" or both. A leader must see and feel the need to achieve a better way to do things (remember, the ultimate goal of affordability is to "make better"). From this need, the purpose, reason, and dream create the vision. Using the eight-step Kotter approach, that leader forms a powerful leadership coalition to investigate the ideas, notions, and concepts that can be leveraged to realize the vision. The collaborative vision is born. Once known, the collaborative vision can be researched, cogitated, pondered, ruminated, discerned, and decided through deep thinking and rational judgment. Once knowledge is grasped, and conviction and consensus is reached, the purpose and vision can be expanded toward development of a design and plan. Methods, means, and measures are identified and determined. The design, plan, and message are developed in order to articulate both verbally and actually what needs to take place, how it will happen, and who will accomplish the results for the final outcome of the first implementation.

Leaders

The leader and the leader core (i.e., key critical leaders in the organization) must serve to spearhead the effort. Leadership teamwork is required to undertake such a herculean and colossal undertaking. It will require numerous changes in the way the work is done, but also a major transformation in the way the culture performs. Collaboration

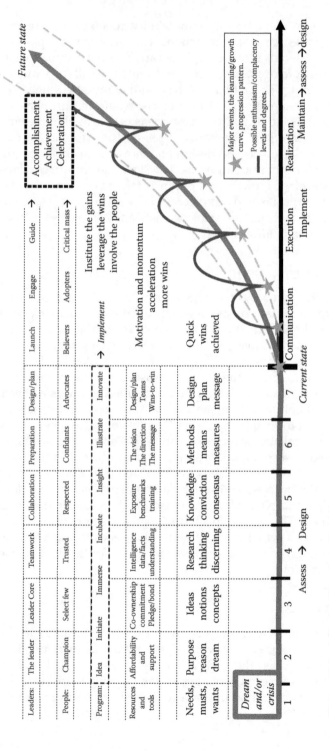

Figure 12.1 Affordability phase line.

will require all leaders to cooperate, participate, engage, and be continually involved. The leadership will be accountable and responsible for ensuring that the plan and design will achieve the purpose, vision, and mission. The strategy and strategic plan for affordability will be the roadmap to success.

People

In the beginning, it is not uncommon for a leader to select an individual, sometimes a top leader, sometimes a middle manager, sometimes a highly respected professional to serve as the champion and super-advocate for the program. After the effort begins, a select few individuals should be chosen to be included in the effort. Such individuals should be trusted, respected, confidants who have the experience, the knowledge, and the wherewithal to participate with the top leaders and be able to bridge between leadership, management, and the people. These advocates are usually the design team that drives the initial implementation activities seeking quick wins to gain momentum. They're involved in designing, planning, and executing the program.

Program

The steps and stages of the program during the assess and design phases follow the creative and innovative process. The program before implementation must embrace the initial idea, initiate creative thinking, be immersed in all possibilities and potentials through an incubation and discovery segment, document insight, illustrate the image and model, and reach innovation yielding implementation.

Resources and tools

Of course, the house of affordability, the triple aim, and the 5P system would be included; however, additional tools, methods, and techniques should be identified and implemented as warranted by the design as it emerges. Early in the first half of the phase line, co-ownership and commitment must be reached and pledged and bonded into the team of people involved. Intelligence, data, facts, and a comprehensive understanding serve for the foundation of what needs to be done. Exposure to examples and benchmarks that contain how others went about attaining success and specific training and development of the team should be accomplished. Once the vision, direction, and message for the program are developed, the design of the solution, the plan of implementation, the identification of "win-to-win," and the preparation of the teams for initial quick wins can be undertaken.

Assess → design

1. A "dream" and/or a "crisis" must emerge for the leader to see and feel the need to pursue affordability and establish the sense of urgency for driving the program initiative. This includes the needs, musts, and wants to be addressed.
2. The leader, with the possible assistance of a designated champion, should hone the idea and determine, other than affordability, what is needed and required. This is where the purpose, reason, and dream are defined and articulated.
3. The leader and the core leaders join forces with a select few individuals from the organization to initiate and immerse themselves in defining the dream more succinctly and attaining co-ownership and commitment, through pledge and bond, to succeed. This is when ideas are developed, notions are defined, and concepts begin to appear.
4. Teamwork is required to guide and shepherd the effort requiring the building and increase of trust and the use of intelligence, data, facts, and understanding to solidify the purpose and vision. It takes research, thinking, and discerning during this period.
5. Now that the team is working together, collaboration should be easily observed. The mental models and team learning should begin generating the sense of what needs to be done and how it is likely to happen. Knowledge, conviction, and consensus should be apparent.
6. Preparation for design and planning using the vision, the determined direction, and the message for consistent communication—this is the realization of methods, means, and measures for the program.
7. Finally, the design and plan are developed and readied for implementation. The end of assess → design occurs upon the delivery of the final design, the initial plan for implementation, the message to be deployed, and the various methods of deployment.

Implement

The implementation is kicked off by communication and deployment of the message, the design, and the plan. The message will describe the why, the design describes the what, and the plan describes the how. Within the plan, the teams identified for the first few quick wins will contain the who, when, and where, and will be prepared and trained to execute. It is the responsibility of the leadership to launch the program with believers engaged and involved to deliver the first few initial victories of success. Once the quick wins are achieved, the execution is launched.

Quick wins create motivation and momentum and are designed to accelerate implementation. Leaders are engaged and should be able to observe adopters begin to come onboard to the new paradigms being created. As gains are made, wins pile up, and people join in the effort, the advancements can be incorporated into the culture and the critical mass for true momentum is being approached. Execution and realization happen after several rounds of prescribed wins are completed, and changes are beginning to be inculcated. During implementation, leadership and management must constantly recognize accomplishment and achievement and find ways to celebrate success.

Maintain

As execution and realization are reached, the standardization and maintenance can commence. Maintenance comes through reinforcement of desired behavior and adjustment of undesired behavior. Reinforcement can be orchestrated through customer compliments, leadership acknowledgement, or trusted team member recognition. As a part of maintenance, an ongoing effort to assess and design the next round should be entertained.

Current state → future state

The "how to" achieve affordability is based on moving an organization's paradigm from its current state to the prescribed and determined future state. To do this successfully, quite a bit of upfront discovery, development, designing, and planning must be completed. The people, from a qualitative perspective, and the things, from a quantitative perspective must all be completed. Since the most challenging part of this transformation is people, there is an affordability quaternity chart that exists to guide the leaders' and teams' efforts.

Affordability quaternity chart for people (see Figure 12.2)

Affordability is about people applying themselves to performance excellence and solving the problems of the organization in pursuit of the purpose, vision, and mission, with shared values, strategy, structure, systems, skills, style and appropriate staffing, integrating value, customer, and cost for continuous improvement and success. To transform an organization from its current state to its desired future state to achieve victory and success, direction, alignment, motivation, communication, and execution all play a part. To accomplish the ultimate goal, we must coach, mentor, and

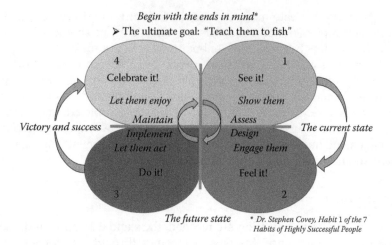

Figure 12.2 The affordability quaternity chart.

teach the people the path to get there. The affordability quaternity chart is that path to get there. Leveraging Dr. Covey's first habit, we must enable and furnish the people with the clear course and route. It's done in four steps and stages:

1. *See it*: Get them to witness what "it" is and what needs to be done. This allows them to assess the current condition and gain an understanding with the possibility of realizing the resources and tools required to solve the problem or attain performance excellence. By showing them an image and reality, it initiates the problem-solving process. As time progresses, the current state should begin to emerge within their awareness.
2. *Feel it*: This happens after they assess and see it. By getting them engaged, potential solutions and resolutions begin to surface in their mind and collective consciousness. A choice for a solution design can be chosen and the crafting of the future state can begin to materialize.
3. *Do it*: The future state being known now, the implementation and actions to institute the solution can occur. This action creates energy, enthusiasm, and motivation. Victory and success is the result.
4. *Celebrate it*: Every victory, no matter how small the success, should be celebrated. The euphoric of "we did it!" allows them to enjoy their achievement. (It should also be noted that setbacks are opportunities for growth and learning ... setbacks take them back to stage 1 to look at it again, stage 2 to experience it again, and stage 3 to try it again with another approach. This is the best and strongest method

of developing people.) The practice of celebration should be maintained and sustained. Creativity and innovation will be required to discovery, identify, and institute new and better forms of reward and recognition.

I've covered the fundamentals of "how to" from the enterprise level, the strategic level, the operational level, and the tactical level. Now let's take a look at three different case examples that demonstrate affordability at work.

Case example: NCR

In 1989, I was working in NCR World Headquarters in Dayton, Ohio, as a strategic planner in the Banking Systems Division. I was asked about my interest in helping open a retail system production facility in Duluth, Georgia. The plant was designed as the first J.I.T. assembly plant within NCR that was devised to consolidate the manufacturing of NCR retail products into one engineering and manufacturing facility. It was a "green field" site, or actually, a pine forest site because the facility was being built on ground that had never been used. It was positioned as NCR's premier site for manufacturing bringing together the retail terminal and canner products produced in Cambridge, Ohio, the retail printers from Ithaca, New York, and the retail peripherals from Dayton, Ohio. This effort was an attempt to inculcate lean thinking and production into the NCR assembly world.

The plant opened in October 1989. Before its first day of operation, quite a bit of preparation had taken place. The surface mount technology circuit board line had been prototyped at a local community college named Gwinnett Tech. The people and resources for the plant from Cambridge, Ithaca, and Dayton were being trained and developed. And, the Lean layout had been architected both from a value stream perspective (see Figure 12.3) and physical floor material flow arrangement (see Figure 12.4). Those first 3 months served to kick-start the system and ramp up for the 1990 production. With this plant in place, most of all NCR's retail system production was designed, developed, and manufactured in one place. The plant was called E&M Atlanta (Engineering and Manufacturing Atlanta) and it contained all of the value-added and value-added support resources (e.g., finance, HR, quality, technical publications, sales and marketing support, etc.) to design, develop, produce, deliver, and 100,000 retail products each year delivered in a just-in-time manner.

In the first year of operation, quite a bit of time was spent in the initial stages of implementation. Much of the training and development was focused on Lean manufacturing principles changing a traditional "push" manufacturing philosophy to a "pull" system. Some of the experienced

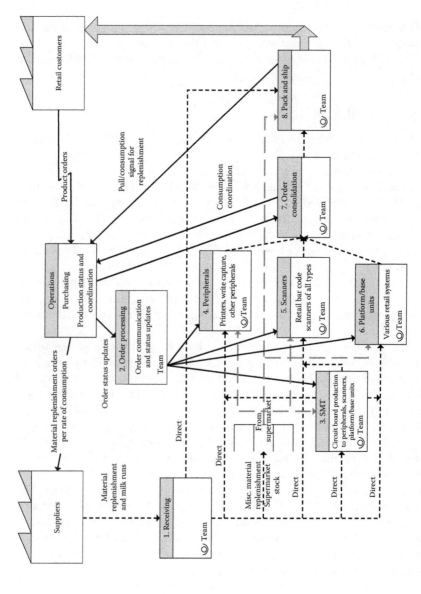

Figure 12.3 NCR, retail systems, value stream map.

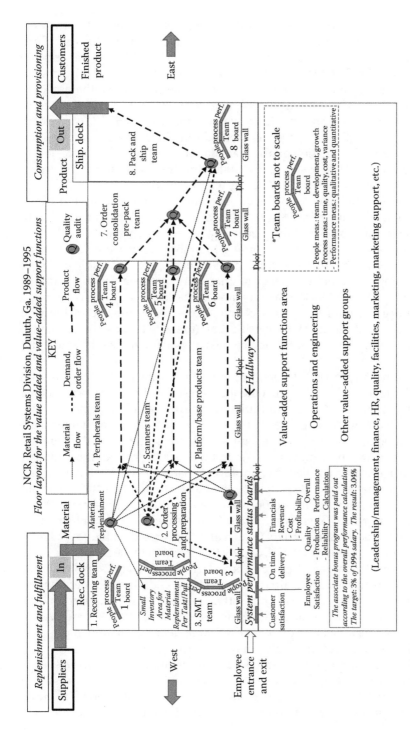

Figure 12.4 NCR retail systems division, Duluth, Georgia, 1989–1995. Floor layout for the value-added and value-added support functions.

personnel not familiar with the new philosophy responded with resistance. The people newly hired locally were much more receptive to Lean methods since this paradigm was not part of their NCR experience being fresh and stimulated to learn and grow. The first year ended with dismal results; the customer satisfaction survey came in a 72.1%, the associate (employee) satisfaction was a miserable 52.0% (about half of the people were satisfied), our on-time delivery performance was 60.2%, our overall system quality (this included a combined functional quality measure and out-of-box for installation quality calculation) was 76.2%, we were able to only produce 72,000 systems, and following the remaining result (especially note the profit loss), in essence, we were failing.

The "dream" and "crisis" were both in place for change and transformation beginning in 1991. The dream of a J.I.T. plant combined with the 1990 crisis results put the pieces in place for staging an "affordability transformation." As 1990 was coming to a close, plans were being developed by leadership to spend time in 1991, developing leadership and people for improving performance and increasing problem-solving efforts to eliminate the roadblocks, barriers, and obstacles to success. The energy at the beginning of the year was palpable. And, by the end of the year we had earned our ISO 9000 certification (first site to do so in the metro-Atlanta area) and our performance measures demonstrated a one-point positive trend.

The next year 1992, leveraging the momentum of 1991, began with the can-do attitude for improvement as a result of 1991's outcome. As the year went on, quite a bit of additional progress was made and the J.I.T. approach to manufacturing and assembly was visibly achieving success. Although the design was simple and straightforward, the demand for systems was now moving closer to the 100,000 mark, and the number of associates (as we called the employees) had reached 1000. About half of the associates were dedicated to the value stream, and the other half dedicated to value stream support. The value stream (see Figure 12.3) was designed to receive customer orders (that created the demand signal to produce), process those orders (which also triggered demand for material provisioning), receive material, deploy that material to the respective areas consuming it, manufacture circuit boards (in the SMT area supplied to the manufacturing/assembly areas), manufacture and assemble peripherals, scanners and platform/base units (often referred to as "retail terminals"), consolidate the orders, and pack/ship the orders to the customer. All of these were designed to work on a rhythm of demand creating a pull system that replenished upon consumption of the products.

Toward the end of 1992, and during the beginning of 1993, the energy and enthusiasm began to wane. Although quite a bit of progress was made from 1989 until now, it became obvious that a new round of initiatives and effort was required to move the organization to the next level of performance. In parallel with other ongoing activities, the development of the people continued, the problem solving of processes continued, and the measurement of performance continued to be communicated and reviewed. That year two major initiatives were employed, one focused on leadership and the other focused on the associates.

The leadership team, who had experienced an outward bound program a couple of years before, to create cooperation, collaboration, and co-ownership of the system among the leadership body, began showing signs of conflict, tension, and diversion. For the first 6 months of 1993, we were finally feeling the pressure of our "new owner," AT&T, who had purchased NCR a couple of years before, and our name had been changed from NCR to AT&T GIS (Global Information Systems) but we still held on to the title E&M Atlanta. In addition, we had hurriedly released a new product, the 7054, which had exhibited numerous operational, functional, and installation problems. We needed a new connection with the "dream" with enough crisis available to rally the leadership's emotional energy to succeed. In August of 1993, I was able to conduct and facilitate a Malcolm Baldrige Assessment with the Leadership Team (my colleagues) to identify areas of opportunity, with a result of 392 out of a possible score of 1000 (if you like, you can go back to Figure 7.5 for details). With our annual strategic planning development coming up in September, and conflict verve caused by merger anxiety, and angst due a product in near fiasco, and a "failing" result on the Baldrige Assessment, the conditions were ripe for another round of improvement.

On the associate side, as 1992 ended and 1993 began, an effort was put in place to expose everyone in the plant to our customer realities. We implemented a program called F.Q.A. (Field Quality Audit), where a team of three associates from the plant would travel to a customer installation site (go to the Gemba) and witness a live installation of our products at various customers around the world. The team of three was composed of one manufacturing associate, one associate from engineering, and one associate from the other value-added support areas (i.e., finance, human resources, quality, marketing support, facilities, administration, etc.). Our goal was to expose every associate to how our products were installed and used and to capture on video the process of installation with actual evidence on video tape (a VHS recorder was provided for the teams to film and archive their trip). Since we had a worldwide customer base, some of the trips provided an opportunity for associates to experience different countries and cultures (i.e., London, Toronto, Tokyo, Sydney, Cape Town). At the end of every quarter of the year, the videos of the trips were edited, combined, and shown to the associates at the beginning of the next quarter. This approach improved our emphasis on customer satisfaction, on-time delivery, and quality. It also served to capture problems or "opportunities to resolve" for the various types of products we produced. Although the products functioned at a 100% operational level when shipped, our overall quality suffered due to installation issues (e.g., wrong cables, missing cables, wrong keyboards, missing or wrong scanners, etc.). The FQAs increased motivation of the workforce, improved on-time results, improved quality by solving installation issues, and increased profitability by lowering cost. In fact, the money being saved was ear-marked to fund an associate performance bonus program for the first time in the history of a company that was founded in 1884 (note: the associate performance bonus did not include the management associates who qualified for the management bonus program!). As 1994 began, a new energy and enthusiasm was abound.

In December 1993 and January 1994, plans were being executed and actions were taking place to once again revitalize the efforts seeking to realize the "dream" of a journey started way back in 1989 and even before. With the associate bonus program rolling out, the FQAs yielding evidence and opportunity, the year was kicked off with an initiative, designed to embrace and leverage the NCR merger reality with AT&T. I was able to get the approval and sponsorship of the senior vice president of the retail division (my boss) to fund an effort to produce an artifact that was designed to bring together all the efforts that had been done from 1989 through 1993. It was easy to gain his commitment because he had the fragment of an idea to produce a music video (quite popular during the early 1990s) to communicate the message of values, customer focus, associate commitment, time, quality, and cost (which by the way were the primary measures of the associate bonus program scheduled to begin January 1994). The last week of the year in 1993 and the first 2 weeks of 1994, the AT&T Creative Resource Group came in to the plant to film and produce the music video, "Our Common Bond," E&M Atlanta, AT&T GIS. The words and lyrics were written by associates, and the video consisted of several songs with all the associates of the plant playing some role in the video. To say the least, it was fantastic. During our 1994 kickoff meetings (We had two. One for the associates working during first shift 07:00–4:00, and one for second shift associates 03:00–12:00. Note: there was a 1-hour hand off period of time between shifts.), the plan for 1994 was discussed, and the video was shown. At the end of the meeting, on the way out, every associate was given their own personal copy of the video.

In addition to the 1994 video, the associate bonus program took place. This was an effort to increase the focus on customer satisfaction, associate satisfaction, on-time, quality, and profit. On the manufacturing and assembly floor, the team performance boards were enhanced and configured to convey performance in three categories: people (the associates, the team, the training, the certification and awards, the team activities), process (the value stream map, process map, and process improvements), performance (qualitative metrics: customer, people, suppliers and quantitative metrics: time, quality, cost). Each of the eight team areas had team boards (see Figure 12.4), and the aggregate of the entire plant performance of the value stream was displayed in a prominent location in the associate flow area of the plant (see system performance status boards, Figure 12.4). The boards were reviewed weekly via a "quality tour" by the plant top line staff of 12 representing each functional area from 10:00 to 12:00 on Fridays (note: with eight areas, only 15 minutes per board were allocated, so the executives had to do a lot of listening to status, updates, and needs requests in that 15-minute period). The system performance board was updated regularly and the overall performance calculation informed the associates of the status and trend for the 1994 bonus payout. It was rather interesting that the summary display of the performance boards, originally designed by associates, resembled a baseball diamond (note: during the early 1990s, the Atlanta Braves were very successful and popular), and when a team met or exceeded the consummate goals for 4 weeks in a row, they earned a run and received recognition and an award for performance. At this point, everyone was interested in our performance, customer satisfaction, associate satisfaction, time, quality, and cost (which included revenue and profit). The target was simple and straightforward. If the plant performed at 100% of the aggregate goals for the year, each associate will receive 3% of their 1994 salary (less than 70% of goal receives 0, 110% of goal or more receives 6%). On the first Friday of 1995, each associate was met at the entrance as they came in the plant to receive a bonus check for the first time in NCR's history for 3.04% of their 1994 salary.

From the beginning in 1989 until 1995, although it started as a "green field" site, it took 5 years of change and transformation to reach a performance level approaching excellence. Although not all the activities and initiatives were included in this story, the patterns of change and transformation, the model for affordability, the affordability phase line, and the methods depicted by the affordability quaternity chart are all apparent. The performance of the organization from the baseline set in 1990 until the outcome of 1994 (see Figure 12.5) illustrates a dramatic victory for the

Measure	Baseline	Outcome
Customer satisfaction	72.1%	85.6%
Associate satisfaction	52.0%	87.0%
Delivery performance	60.2 %	93.0%
Overall system quality	76.2%	99.7%*
Production capacity	72,000	98,000
Development cycle time	24-36 months	12-18 months
Revenue	$220MM	$282MM
Profit	$2MM	$5MM

Figure 12.5 NCR: affordability.

people of the organization. This is the example and archetype that jump-started my understanding of the theory of affordability.

Case example: SASI

In December of 1997, after having incorporated my consulting business in September of 1995, I was invited in to Bristol PA to meet with the owners of a company called Store Automated Systems, Inc. (S.A.S.I.). The four owners had been in business several years after merging two companies in the late 1980s to form a point-of-sale systems company offering PC-based P.O.S. systems to retailers (e.g., Rite Aid Pharmacy, Pep Boys, etc.). Their dilemma stemmed from what termed the "worst part of their organization," the technical help desk. They wanted me to "fix" the help desk. One of the owners, Jon Hayward, the chief scientist and systems VP, had the help desk reporting to him. He was the leader, owner, and champion of the effort to "fix the help desk." I considered him a genius when it came to software, systems, and retail transaction products. However, although the help desk was under his span of control, the priority of his attention had to be applied to the products, the software, and integrated customer solution. Working way more than 40 hours and 5 days each week, he knew the help desk suffered, not due to his availability, but as much due to the fact that none of the owners wanted to deal with it.

In January 1998, I visited the help desk first hand (i.e., went to the Gemba). My first experience was met with fear, anxiety, and trepidation. The two managers and the people thought I was there to identify who would need to be laid off. No one explained to them I was there to support their efforts to solve problems, improve processes, increase performance, "teach the fishers to fish," and make better. The current state of the help desk existed as a "cube farm," where each employee had an office cube, a computer, a phone, and little else (see Figure 12.6). It was a three-shift operation, with a first and second shift manager/lead including seven associates for each shift, and a "skeletal" third shift consisting of two associates. Their work involved fielding technical calls from customers and resolving field problems over the phone. That first day was revealing; the flow of the calls and the information was chaotic and random, it also was not designed to be smooth, they pushed calls to problem hold areas, and they didn't have any form of performance metrics (i.e., they didn't know how many calls they handled in a day, week, month, or year, they didn't have documentation at their desks, they didn't have systems to test solutions, they didn't have access to engineers, they didn't even have standard training on the systems they were supported). In fact, toward the end of that first week of assessment, I was told (by the managers and the people) the only way you can get out of this place is to quit the company or die. It was known as the worst place to work in the company.

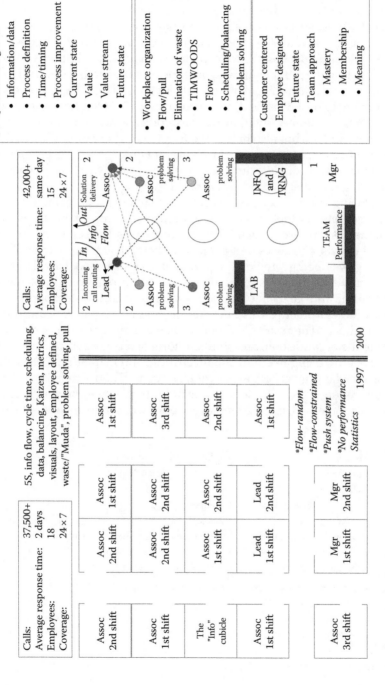

Figure 12.6 SASI help desk: technical service and support.

After preparation and planning (assess → design), the first event and activity was performed to begin the motivation and standardization. It consisted of two components: 5S and organization performance. We were part of a Tuesday and part of a Wednesday performing a 5S in every cube. As a result, we were able to begin to architect what a good help desk cube looks like and operates like to serve the customers. In addition, we spent a week noting the number of calls, the type of calls, and the time duration it took to resolve the problem of each call. By the end Q1 1998, it was realized that the help desk was fielding approximately 37,500 calls annually, with an average customer solution response of 2 days (note: half of them in two days or less, but also half of them in two days or more). Most of 1998 was spent in rapid improvement events and quick wins and performance improvement through techniques involving flow, cycle time, scheduling, balancing, visuals, metrics, and problem solving.

During 1999, after the success and momentum gained in 1998, advanced efforts took place to reconfigure and implement an associate designed layout and system. The customer was the center of all value-added efforts, processes (not people) we being fixed, support from engineering was being provided, equipment was provided for a lab to test solution, teamwork was incorporated, flow and pull were implemented, scheduling and balancing techniques were utilized, a system performance measurement practice was put in place, and an employee "future state" was realized. The results and outcomes were more than apparent to the rest of the organization, and similar efforts of change and transformation were begun in engineering, manufacturing, and marketing (by starting a strategic planning effort). In Figure 12.6, you will notice that the 2000 floor layout reduced the help desk footprint by half, while the ability to handle calls increased, but the employee count went from 18 to 15. The three

Measure	Baseline	Outcome	
Timeframe	Dec 1997	Jan 2000	
Help desk	18	15*	
Service response average	48 hours	Same day	
Service support cost	$1,080,000	$675,000	(37.5% reduction)
Product development cycle	20-36 months	12-18 months	(twice as fast!)
Strategic plan	None	5-year plan	
Company valuation	$13.4MM	$52.0MM	

- The three best associates were promoted to other areas.
  - Engineering
  - Manufacturing
  - Market and planning

Figure 12.7 SASI: affordability.

"other employees" were actually promoted to engineering, manufacturing, and the new marketing and strategic planning area to deploy the improvement methods to other parts of the organization.

The overall outcome and results over a 2-year period is a testimonial for the power of affordability. Performance went up, cost went down, speed increased, strategy and strategic planning was instituted, and a company that was valued at around $13M in 1997, was now worth $52M in 2000. Above all, people were recognized, rewarded, and celebrated for achievement and accomplishment (Figure 12.7).

Case example: Northrop Grumman

In my final "how to" case study, I feature the project that actually uncovered the name "affordability" to this theory that combined customers, people, suppliers, and partners, with process, performance, and profitability, for an environment designed to "make better" by integrating value, customer, and cost for continuous improvement.

With the vision and guidance of two senior leaders, Mr. George Vardoulakis, senior VP, and Mr. Dave Armbruster, program director, the partnership of the IIE and Dr. Beth Cudney, the participation of numerous key and critical employees (and the eventual engagement of thousands of employees by the end of 2009), the affordability program at Northrop Grumman El Segundo CA was begun at the F/A-18 Super Hornet plant in 2007. This effort was primarily positioned to be a Lean initiative. My partner, Dr. Beth Cudney, and I didn't know going in that Lean efforts had failed in 2000, 2002, and 2004. And, what was more enlightening, each time Lean was "tried" it failed due to overwhelming resistance and rejection. The employees of the plant referred to Lean as an acronym L.E.A.N., Less Employees At Northrop. At that point in time, we knew Lean would likely fail again. However, the vision put forth by George and Dave was much more than Lean. We eventually named the program "affordability." This solidified the name for the theory of affordability.

By 2007, the U.S. Navy and the United States Marine Corps, the customers for the F/A-18 Super Hornet, had been involved in the wars in Afghanistan and Iraq for several years. The demand for replacement products was on the rise. Northrop Grumman owned 60% of the statement of work for Super Hornet production with Boeing being the parent company, the El Segundo plant sent a jet, nearly two-thirds completed to St. Louis where GE installed the engines, Raytheon completed the cockpit electronics, Boeing finished the plane and sold it to the U.S. government. At the time, a 42-unit capacity existed as a result of the 2006 production. The cycle time for manufacturing was on a rhythm of 1 every 5.5 days with a good quality level at a price of $55,000,000.00 per plane.

With the increase in U.S. demand and the opportunity to sell the product to other nations, a new price point target of $49,900,000 was set as a result of international customer requirements. These new requirements put in motion the need for affordability.

Over a timeframe of 3 years (2007, 2008, 2009), the affordability program was rolled out to address the new requirements and improve production performance. The first employees engaged in the effort consisted of about 18 individuals from leadership and "the floor" (manufacturing employees), which included George, Dave, some of the top-line staff, and around nine manufacturing employees. Three initial quick win projects were identified with George and Dave serving as champion and participant to ensure success. Every 2 months thereafter, three wins were identified and completed to maintain momentum. Eventually, by the end of 2009, all plant employees were involved and by August of 2010 the affordability program was yielding fantastic results (Figure 12.8).

With the outcome of the Northrop Grumman project, and the last 5 years from 2010 until now, I have been developing and refining the theory of affordability in order to communicate how to go about achieving affordability. In order to be able to incorporate affordability for continuous improvement, I suggest you use this book as a guide. What follows is a checklist that assists in maintaining focus and creating a comprehensive solution to attain affordability:

- The affordability basics
- The value
- The customers and markets
- The costs
- Faster and better
- Leadership
- Change and transformation
- Creativity and innovation
- People
- Process
- Performance
- Purpose and direction
- Values and ethics
- Culture and success
- Partnerships and relationships
- Adaptability and flexibility
- Community and environment
- Learning and growth
- Organization and governance
- Financials and prosperity
- Accomplishment, achievement, and progress

The affordability program 2007–2009

Performance metrics/measures	Baseline 2006	Results 2010
• Production capability	42	62
• Cycle time	1/5.5 days	1/4.0 days
• Quality	5σ	6σ
• Base price	$55,000,000	$49,900,000 ($5.1M reduction)
• Customer base	United States	United States, Australia, and others

F/A–18
"Super Hornet"

*** Note: scheduled plant closing 08/2010 → plant life extended to 2020
In August 2010, the U.S. government purchased 124 planes

Figure 12.8 Northrop Grumman: affordability.

How to do it

Assess the current state

- Research
- Understand
- Map and model

Design the future state and create the plan

- Plan and schedule
- Set direction and align the resources
- Develop the message

Implement the design and plan

- Purpose, vision, direction
- Deploy the program
- Leadership
- People and teams
- Resources/tools
- Design/plan
- Quick wins
- Communication
- Execution
- Celebration

Maintain the momentum and sustain the improvements

- Continuously measure and monitor performance
- Adjust and upgrade

- Innovate and improve
- Assess
- Design
- Continue the effort with constancy of purpose and continuous improvement

Now do it! Make better!

chapter thirteen

The affordability challenge

Now is the time to use Affordability to make all things better!

—Paul W. Odomirok

Create a constancy of purpose for improving products and services (and systems) and adopt the new philosophy are the first two of Dr. Deming's 14 points. "Create a sense of urgency" is step one of Dr. John Kotter's eight-step transformation model. Model the way, inspire a shared vision, challenge the process, enable others to act, and encourage the heart are the five practices of exemplary leadership behaviors identified by Dr. Jim Kouzes and Dr. Barry Posner. Set direction, align the resources, and motivate the people are what leaders really do according to Dr. John Kotter. This is the essence of the affordability challenge. All of the recommendations of these gurus come into play when integrating value, customer, and cost for continuous improvement. In other words, go out and "Make all things better!"

Organizations cannot survive on price, profit, and cost alone. Too often, leaders set a major goal of reducing cost without balancing this goal with increasing value and improving customer offerings. Cost alone is a shortsighted objective. Without integrating value and customer with cost, an imbalance exists, and the ultimate result is most likely failure. Likewise, focusing only on the customer, while value and cost suffer, failure also likely is probable. And, it consistently follows: concentrating on only value, and not customer, nor cost, eventually results in failure as well. The affordability challenge is about all three or nothing. A win-win-win strategy.

Over the years, I've observed company after company, institution after institution crash and burn using a monolithic focus and sometimes on an objective other than value, customer, and cost. I cannot recall any organization surviving on cost alone, nor customer alone, nor value alone, nor on any other disparate goal. The companies created in the early twentieth century that survived into the twenty-first century demonstrated the tenets of affordability. Some of those companies come from the automobile industry. Both Toyota and Ford are good examples of companies that continually provide increased value, meet customer requirements, and focus on keeping costs as low as possible. In addition to affordability, the companies demonstrating staying power had many other affordability elements, such as clear purpose and direction, strong culture and people consciousness, durable customer and supplier relationships, clarity of value and competency, and a long-term attitude toward business, community, and environment.

The oldest 10 publicly traded companies (Lorillard Tobacco, 1760; Baker's, 1765; Ames True Temper, 1774; Bowne, 1775; Bank of New York Melon, 1784; Cigna, 1792; State Street, 1792; Jim Beam, 1795; JPMorganChase, 1799; DuPont, 1802) have demonstrated the triple aim for focus for longevity. In fact, there were at least 14 family-created and privately owned businesses (Shirley Plantation, 1613; Tuttle's Red Barn, 1632–1633; Field View Farm, 1639; Barker's Farm, 1642; Seaside Inn, 1667; White Horse Tavern, 1673; Saunderskill Farm, 1680; Towle Silversmiths, 1690; The John Stevens Shop Stone Carving,1705; Orchards of Concklin, 1711; Smiling Hill Farm, 1720s; Lakeside Mills, 1736; WD Cowels Lumber, 1741; Caswell-Massey Perfume, 1752) that preceded the oldest 10 public companies who are still in business today.

If you desire longevity, if you want to increase your competitiveness, if you yearn to grow your customer base, if you wish to increase your market share, if you want to stay in business, incorporate affordability for continuous improvement, focus on

- Affordability as a strategic framework
- Increasing value
- Improving customer and market offerings
- Reducing cost
- Providing products and service faster and better
- Instituting leadership
- Preparing and planning for change and transformation
- Instituting creativity and innovation
- Growing and motivating people
- Challenging and improving processes
- Instituting performance metrics and measures and rewarding and celebrating success
- Communicating purpose and direction
- Establishing and nurturing resilient values and ethics
- Maintaining a strong culture and emphasize Success
- Maintain solid partnerships and relationships both downstream and upstream
- Being adaptable and flexible
- Supporting and investing in the community and the environment
- Encouraging and supporting learning and growth practices and behaviors
- Establishing a clear and solid organization with sound governance constructs
- Following conservative financial practices and celebrating prosperity

In addition and in conjunction with these focal points, there are several document traits that are prevalent within long-living organizations that practice affordability and continuous improvement:

Leadership development: Leadership is vital to success, and leadership development and succession planning is essential. Affordability companies develop their current and future leaders and plan for succession. Purpose, vision, mission, and service require continuity of leadership and succession plan execution.

Relationships: Affordability organizations have a strong ongoing focus on their relationships, both downstream and upstream. Affordability organizations put much more emphasis on their relationships with suppliers, customers, and local communities than other organizations. Successful affordability organizations understand that they have an obligation and responsibility to support their community as well as their stakeholders and customers.

Intelligent change and transformation: Affordability organizations are adaptable and flexible and open to change, but at the right pace. Affordability organizations with their strong culture and unique identity focus on their entity tradition and improve what they realize as their core strengths. But they are ever cognizant that change is inevitable and embrace their own evolution. When large-scale and significant change is necessary, they take a long time to prepare, plan, and implement change, and they leverage their creativity and they practice purposeful innovation.

Conservative financials: Affordability organizations use and utilize conservative financial practices. They are excellent planners that pride themselves in having conservative fiscal practices. Affordability organizations prefer profitability over sales volume and are reluctant to borrow money, opting instead to grow conservatively with cash in hand from realized cost reductions, and additional profits through improved products and services.

Purpose and values: Affordability organizations have a strong sense of purpose and values. They weigh decisions based upon those values and build leaderships that can align themselves to the values and ethics that the company was founded upon. They have the ability to move into new markets and new customer niches without abandoning core tenets and quickly align the new ventures with the established core principles.

These focal points and traits are anchored within the heart of affordability. They are the customs and core habits for increasing value, exceeding customer expectations, and reducing cost. They provide the objective areas for attaining growth and achieving durability.

My challenge for you: Institute affordability.

Index